Energy for Keeps

Creating Clean Electricity from Renewable Resources

EXPANDED 3rd EDITION

Marilyn Nemzer
Deborah Page
Anna Carter

Illustrated by
Will Suckow

Energy Education Group
Tiburon, California

Energy for Keeps, 3rd Edition:
Creating Clean Electricity from Renewable Resources

©2010 Energy Education Group
ISBN: 9780974476551
Library of Congress Control Number: 2010924188

Library of Congress Cataloguing-in-Publication Data for earlier editions:
Nemzer, Marilyn L., 1942-
  Energy for keeps: electricity from renewable energy: an illustrated guide for everyone who uses electricity
Marilyn L. Nemzer; Deborah S. Page; Anna K. Carter; Will Suckow, illustrator.
      p. cm.
  Includes bibliographical references and index.
  ISBN 0-9744765-2-8 (pbk.)
1. Renewable energy sources. 2. Electric power production. I. Nemzer, Marilyn, 1942- II. Title.
  TJ808.P326 2005
  333.793'2—dc22
  2005044259

Published by the Energy Education Group (aka Educators for the Environment), a division of The California Study, Inc., a 501(c)(3) nonprofit organization. No portion of this book may be reproduced for publication or for sale without the written permission of the Energy Education Group.

664 Hilary Drive
Tiburon, California 94920
415.435.4574
energyforkeeps@aol.com
www.energyforkeeps.org

Manufactured by Friesens Corporation, Altona, MB, Canada, in May 2010; job #55255.
Printed on 100% pcw recycled paper.

LEGAL NOTICE
The first edition of *Energy for Keeps* was prepared as a result of work sponsored by the California Energy Commission. This publication does not necessarily represent the views of the Energy Commission, its employees, or the State of California. The Commission, the State of California, its employees, contractors, and subcontractors make no warranty, express or implied, and assume no legal liability for the information in this book; nor does any party represent that the use of this information will not infringe upon privately owned rights.

COVER PHOTOGRAPHS
Background: *Sky of July*, courtesy Constantin Jurcut, UK
Biomass: Courtesy of PhotoXpress.com
Geothermal: Fumaroles in the Philippines, courtesy of Geothermal Education Office, Tiburon, CA
Hydro: *Waterfall near Shirahone Onsen, Japan,* courtesy of Ian Monroe, Mill Valley, CA
Ocean: *Sun Curl*, courtesy of Clark Little Photography, HI
Solar: Photo by Wayne Gretz, courtesy of National Renewable Energy Laboratory, Golden, CO
Wind: Courtesy of PhotoXpress.com

Editor and co-author: Marilyn Nemzer, Energy Education Group, Tiburon, CA
Lead writer and co-author: Deborah Page, Page One Productions, Claremont, CA
Technical editor and co-author: Anna Carter, Reno, NV
Technical editor and contributor, 3rd edition: Peter Asmus, Stinson Beach, CA
Photo editor: Nikki Nemzer, Burbank, CA
Illustration: Will Suckow Illustration, Sacramento, CA
Design: Barbara Geisler Design, Sausalito, CA
Advisor: Kenneth Press Nemzer, Tiburon, CA

TABLE OF CONTENTS

About this Publication . . . . *iii*
Acknowledgments . . . . *iv*
Benjamin Franklin: An Inspirational Figure . . . . *vii*

1 A BRIEF HISTORY OF ENERGY
How our use of energy has changed over time 17

2 ENERGY AND ELECTRICITY
How we produce and deliver most of our electricity 27

3 ENERGY SOURCES FOR ELECTRICITY GENERATION
How we use different energy sources to produce electricity . . . 33
RENEWABLE ENERGY RESOURCES
    Biomass 39
    Geothermal 49
    Hydropower 61
    Ocean 73
    Solar 83
    Wind 95
THE RENEWABLE AND NONRENEWABLE RESOURCE
    Hydrogen 107
NONRENEWABLE ENERGY RESOURCES
    Fossil Fuels 117
    Nuclear 125

4 ENERGY, HEALTH, AND THE ENVIRONMENT
How energy choices affect our health and the environment . 133

5 ENERGY MANAGEMENT STRATEGIES AND ENERGY POLICY
How energy decisions affect our lives 143

APPENDIX
Energy Timeline 159
Glossary 165
Additional Information Resources 177

INDEX 185

THE AUTHORS 191

ABOUT THIS PUBLICATION

Energy for Keeps offers an introduction to renewable energy for everyone who uses electricity — from students to energy policy makers. This book helps readers of all ages understand the energy issues that loom large in our daily news.

With clear language and engaging illustrations, *Energy for Keeps* covers all renewable energy sources, the science of electricity generation, energy history, environmental considerations, and energy management and efficiency.

Energy for Keeps explains both renewable and nonrenewable energy resources, with an emphasis on renewables. It does not promote a particular technology. To ensure the book's accuracy, the authors interviewed — and had drafts reviewed by — experts from utilities, universities, state and federal agencies, national laboratories, power suppliers and industry. (See Acknowledgments, page iv.)

Aimed at furthering energy literacy for the general public, *Energy for Keeps* also serves as a great text for students of many ages. On the *Energy for Keeps* website, educators will find student activities and other supplementary information that may be downloaded free.

Earlier editions of *Energy for Keeps* received the Interstate Renewable Energy Council's 2004 Innovation Award and a 2006 Green Power Leadership Award from the U.S. Environmental Protection Agency, the U.S. Department of Energy, and the Center for Resource Solutions.

Utilities, government agencies, energy companies, and other entities may want to provide *Energy for Keeps* to their non-technical staff, or to public libraries or schools. For details please contact the Energy Education Group, **www.energyforkeeps.org**.

THE ENERGY EDUCATION GROUP

The Energy Education Group is a division of The California Study, Inc., a nonprofit 501(c)(3) organization based in Tiburon, California. Its expertise is in renewable energy education with a focus on power generation. Its goal is to help people understand where our electricity comes from and how energy choices affect our lives, our environment, and future generations.

ACKNOWLEDGMENTS

The Energy Education Group is grateful to the sponsors of *Energy for Keeps* and to the reviewers and contributors who so generously gave their time to ensure the accuracy of this book.

SPONSORS
Thank you to the California Energy Commission for the initial funding of *Energy for Keeps*. Thank you also to our other major sponsors: Foundation for Water and Energy Education, Bonneville Power Administration, Ormat Technologies, Inc., U.S. Geothermal, Inc., Calpine Corporation, Northern California Power Agency, Southern California Edison Company, Fuji Electric Co. Ltd., Oregon Department of Energy, Mid-American Energy Holdings Co., Big Bend Electric Cooperative, Inc., Baker Hughes Geothermal Operations, and Mammoth Pacific L.P. And thanks to Scott McInnis and the other generous individuals who contributed financially. We also want to acknowledge the benefits this book derived from previous publications by the authors under sponsorship from the U.S. Department of Energy (Geothermal Technologies Program).

REVIEWERS AND CONTRIBUTORS
During the development, writing, and editing of *Energy for Keeps*, we called upon many technical and educational experts. We are deeply grateful to all of them, and we thank especially the following:

Rodney Aho, *Bonneville Power Administration, Portland, OR*
Donald Aitken, *Donald Aitken Associates, Berkeley, CA*
Bill Andrews, *California Department of Education, Sacramento, CA*
Roger Bedard, *Electric Power Research Institute, Palo Alto, CA*
Betz Bornstein, *power generation consultant, Piedmont, CA*
Pat Byrne, *Sacramento Municipal Utility District, Sacramento, CA*
Barbara Byron, *California Energy Commission, Sacramento, CA*
Clyde Carpenter, *Foundation for Water and Energy Education, Spokane, WA*
Rebecca Clark, *Bonneville Power Administration, Portland, OR*
Ruth Cox, *U.S. Fuel Cell Council, Washington, D.C.*
Jeff Deyette, *Union of Concerned Scientists, Cambridge, MA*
Lynette Esternon, *California Energy Commission, Sacramento, CA*
LaVonne Ewing, *PixyJack Press, Inc., Masonville, CO*

ACKNOWLEDGMENTS

Phyllis Evans, *Bonneville Power Administration, Portland, OR*
Susanne Garfield-Jones, *California Energy Commission, Sacramento, CA*
Karl Gawell, *Geothermal Energy Association, Washington, D.C.*
Barbara George, *Women's Energy Matters, Fairfax, CA*
Dennis Gilles, *Calpine Corporation, San Jose, CA*
Mitchel Gorski, *Covanta Energy, Fairfield, NJ*
Jim Green, *National Renewable Energy Laboratory, Golden, CO*
George Hagerman, *Virginia Tech Alexandria Research Institute, Alexandria, VA*
Marilyn Hempel, *environmental educator, Redlands, CA*
Susan Hodgson, *energy historian, Sacramento, CA*
Jacqui Hoover, *Natural Energy Laboratory of Hawaii Authority, Kailua-Kona, HI*
Ron Horstman, *Western Area Power Administration, Lakewood, CO*
Cynthia Howell, *National Renewable Energy Laboratory, Golden, CO*
Ronald Ishii, *Alternative Energy Systems Consulting, Inc. Carlsbad, CA*
Ellen Jacobson, *University of Nevada College of Engineering, Reno, NV*
Steve Jolley, *Wheelabrator Shasta Energy Company, Anderson, CA*
Tony Jones, *OceanUS Consulting, San Francisco, CA*
Doug Jung, *Two-Phase Engineering, Santa Rosa, CA*
David Kay, *Southern California Edison Company, Rosemead, CA*
Felix Killar, *Nuclear Energy Institute, Washington, D.C.*
Tom Kimbis, *The Solar Foundation, Washington, D.C.*
Bob King, *Ashuelot River Hydro, Inc., Keene, NH*
Lauri Knox, *Vortex International, Inc., Golden, CO*
Matt Kuhn, *National Renewable Energy Laboratory, Golden, CO*
Joe LaFleur, *geologist, Springfield, OR*
Diane Lear, *National Hydropower Association, Washington, D.C.*
Chris Lee, *Pondre School District, Fort Collins, CO*
Peter Lehman, *Humboldt State University, Arcata, CA*
Carl Levesque, *American Wind Energy Association, D.C.*
Marcelo Lippmann, *Lawrence Berkeley Laboratories, Berkeley, CA*
Debra Malin, *Bonneville Power Administration, Portland, OR*
Randy Manion, *Western Area Power Administration, Lakewood, CO*
Valencia McClure, *Exelon Power, Kennett Square, PA*
Michael McCormick, *Johns Hopkins University, Baltimore, MD*
Roy Mink, *U.S. Dept. of Energy Geothermal Program, Washington, D.C.*

(continued)

ACKNOWLEDGMENTS (continued)

Colin Murchie, *Solar Energy Industries Association, Washington, D.C.*
M. Dennis Mynatt, *Tennessee Valley Authority, Knoxville, TN*
Dan Neary, *U.S. Department of Agriculture, Flagstaff, AZ*
Bob Neilson, *Idaho National Laboratory, Idaho Falls, ID*
Craig Nesbit, *Exelon Generation, Warrenville, IL*
Tom Osborn, *Bonneville Power Administration, Portland, OR*
Terrin Pearson, *Bonneville Power Administration, Portland, OR*
Christine Real de Azua, *American Wind Energy Association, Washington, D.C.*
Marshall Reed, *U.S. Geological Survey, Palo Alto, CA*
Hal Post, *Sandia National Laboratory, Albuquerque, NM*
Carl Rivkin, *National Renewable Energy Laboratory, Golden, CO*
John Romero, *U.S. Department of the Interior, Camarillo, CA*
Bob Rose, *U.S. Fuel Cell Council, Washington, D.C.*
Julie Scanlin, *University of Idaho, Boise, ID*
Phil Shepherd, *National Renewable Energy Laboratory, Denver, CA*
Karen Skinner, *Novato Unified School District, Novato, CA*
Bill Smith, *Northern California Power Agency, Middletown, CA*
Arthur Soinski, *California Energy Commission, Sacramento, CA*
Kevin Starr, *California State Librarian (Emeritus), San Francisco, CA*
Ron Stimmel, *American Wind Energy Association, Washington, D.C.*
Jason Venetoulis, *sustainability expert, Claremont, CA*
Mira Vowles, *Bonneville Power Administration, Portland, OR*
Michael Van Brunt, *Covanta Energy, Fairfield, NJ*
Kit Warne, *(ret.) Pacific Gas and Electric Co., Corte Madera, CA*
Judith Wilson, *editorial consultant, Tiburon, CA*
Cindy Wyckoff, *Biomass Energy Resource Center, Montpelier, VT*
Dora Yen-Nakafuji, *California Energy Commission, Sacramento, CA*

It was an honor to work with the experts named above and with my
co-authors, Deborah Page and Anna Carter; Will Suckow, illustrator
extraordinaire; Barbara Geisler, designer; and Nikki Nemzer, photo
editor.

Lastly, we thank Kenneth Press Nemzer for his nonstop support and
good advice on all aspects of the publication.

Marilyn Levin Nemzer, Editor
May 2010

BENJAMIN FRANKLIN: AN INSPIRATIONAL FIGURE

BENJAMIN FRANKLIN HOSTS THE PAGES of *Energy for Keeps*. We chose him not only for his contributions to the field of electricity, but also because he always sought, through hard work and ingenuity, to understand the world around him and to make a positive impact on it.

BENJAMIN FRANKLIN: 1706 – 1790

The best-known story about Ben Franklin is that he experimented with electricity by flying a kite in a raging lightning storm. In reality he did not stand out in a storm (a soaking wet string could have made this experiment fatal), nor was he trying to have lightning actually strike his kite.

Ben had been studying electricity. He had correctly proposed that the sparks resulting from what we now call static electricity — an object of great fascination at that time — were due to excess electrical charges building up in an object and then leaping, or discharging, to an object of lesser charge. He speculated that thunderclouds, too, could build up excess electrical charges and that lightning was the discharge from the cloud to the ground (or other object such as a tree or house). He thought he could prove this theory by flying a kite just before a storm began (before the thunder, lightning and rain started), hoping to draw "fire" (electrical charge) out of the clouds.

So, one day in June of 1752, when a storm was brewing, he tested his idea. He placed a metal wire on a kite's upper tip and tied a metal key to the bottom of the kite string. Standing in a shed as protection from the potential downpour, he flew his kite up into the dark clouds. When the fibers on his kite string began standing up, he gently touched the key and must have been pleased to feel an electrical charge. His experiment confirmed that thunderclouds generate static electricity. He also correctly concluded that lightning results from the build-up and discharge of excess electrical charges.

(continued)

BENJAMIN FRANKLIN (continued)

Ben was not just an avidly curious scientist, but also a writer, a publisher, an inventor, a civic leader, and a statesman. He had his own print shop where he wrote and produced a newspaper and an annual almanac, among other publications. His many inventions include the lightning rod, bifocal glasses, the Franklin stove (a free-standing fireplace), and the odometer (which measures mileage). He began the nation's first lending library and the first fire department. He was Postmaster General of the American colonies. He contributed significantly to the writing of the Declaration of Independence and worked for the abolition of slavery. To top it off, his close diplomatic and scientific ties with Europe influenced France to support the colonial Americans during the Revolutionary War.

For his contributions to science and society, we are pleased to honor Ben Franklin as the host of *Energy for Keeps.*

NEVER A DULL MOMENT

Life with Ben must have been pretty interesting. Imagine living with him while he was testing his new invention, the lightning rod. A metal rod on the roof attracted lightning, which traveled safely to the ground through a wire, sparing the house from fire. In one experiment, he threaded the wire right through the inside of his own house along the staircase banister. One stormy night the family awoke to the sound of bells clanging wildly. It turned out that Ben had attached metal bells to the wire along the banister, so that he would be alerted when electricity passed through to the ground.

Energy for Keeps

CHAPTER 1

A BRIEF HISTORY OF ENERGY
How our use of energy has changed over time

TERMS IN GLOSSARY

alternating current (AC)
A.D.
alloy
alternator
B.C.
blast furnace
charcoal
coke
combustion
direct current (DC)
dynamo
electromagnetism
energy conservation
fossil fuel
generator
geothermal
heat engine
hydropower
industrial
Industrial Revolution
internal combustion engine
manufacture
mass
mass produced
medieval
organic
passive solar
power
smelt
static electricity
Stirling engine
telegraph
textile
town gas
transmit
voltage
wet-cell battery

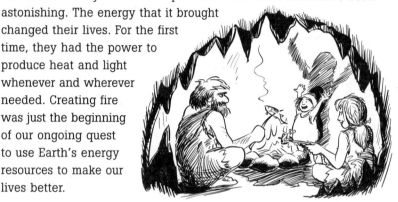

ANYONE WHO'S EVER LIT A CANDLE knows that making fire is as easy as striking a match. But for our earliest ancestors, the ability to create a spark and build a fire must have been astonishing. The energy that it brought changed their lives. For the first time, they had the power to produce heat and light whenever and wherever needed. Creating fire was just the beginning of our ongoing quest to use Earth's energy resources to make our lives better.

OUR FIRST ENERGY SOURCES

For most of the history of humankind wood was the mainstay of life — for shelter, for transportation on land and on water, and as a source of energy to burn for heat and light. Besides using wood and their own muscles, people took advantage of the energy that the sun, wind, running water, hot springs and even animals could provide — to do work, to travel, and for recreation.

Ancient civilizations advanced the use of energy resources. Around 3,500 B.C. (about 5,500 years ago) Egyptians made the earliest known sailboats, harnessing the power of the wind to travel faster and further, while increasing trade with neighboring lands. By 500 B.C. Greeks were building what we now call "passive solar" homes to take better advantage of the sun's light and warmth. And by 85 B.C. Romans were enjoying baths heated with water from geothermal hot springs.

Around the same time, the Greeks made advances in use of running water. They developed waterwheels to grind grain, a task previously done by hand or with animal power. And by 640 A.D., in what is now Iran, the Persians had also found a new way to grind grain, using mills with large wooden blades to capture wind power. Europeans adopted the idea and used modified versions of these windmills throughout medieval times. Next to wood, wind ranked as a prime source of energy.

But wood remained the most-used energy resource. In the 1300s Germans built the first blast furnaces to burn wood at extremely high temperatures, allowing them to produce large quantities of iron. During the next few centuries much of Europe's forested area was logged for the production of iron and the building of ships.

COAL POWERS INDUSTRY

Although people burned coal — a fossil fuel — for heat at least as early as the first century A.D., it took more than a thousand years for coal to become a dominant source of energy. By the late 1600s coal had become more popular than wood in England. In fact, the British had an abundance of coal. But they had flooding problems deep in the coal mines due to groundwater flows. They needed a way to pump out the water.

Fortunately, in 1698, Thomas Savery invented one of the earliest workable steam engines. When attached to a water pump, this engine largely solved the flooding problem. Blasts of steam from water boiled by burning coal kept the engine working whenever the pump was needed.

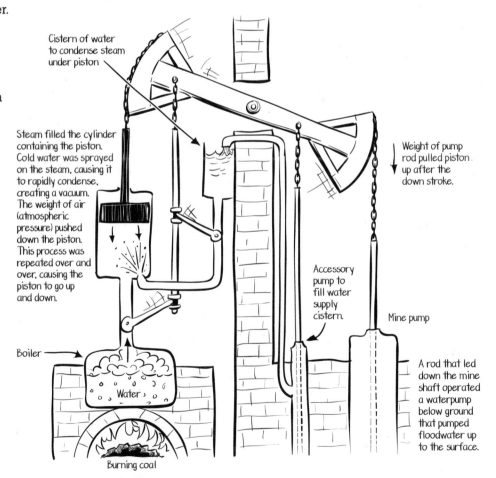

Cistern of water to condense steam under piston

Steam filled the cylinder containing the piston. Cold water was sprayed on the steam, causing it to rapidly condense, creating a vacuum. The weight of air (atmospheric pressure) pushed down the piston. This process was repeated over and over, causing the piston to go up and down.

Weight of pump rod pulled piston up after the down stroke.

Accessory pump to fill water supply cistern.

Mine pump

Boiler

Water

Burning coal

A rod that led down the mine shaft operated a waterpump below ground that pumped floodwater up to the surface.

An early steam engine pumps water from a coal mine.

**A steam-driven cotton mill in the 1700s,
where machinery was powered by steam
engines located in a separate room**

The textile industry was also flourishing in England during this time. It, too, saw revolutionary change, as inventors developed coal-fired steam-driven machines to spin yarn and weave cloth faster and cheaper than the waterwheels that powered early factories. These early steam engines could do the work of tens, even hundreds, of human beings. (Waterwheels remained popular, however, and people continued to use them — as they had for centuries — for such work as lifting water for irrigation and rotating huge stones to grind grain.)

THE INDUSTRIAL REVOLUTION STEAMS AHEAD

By the early 1700s industry was booming, with improved steam engines providing power for machinery to process raw materials and to manufacture products. The new engines required ever-greater amounts of coal to heat water for steam, so the coal-mining industry was a big business. New opportunities beckoned, and lifelong country dwellers migrated to places where they could take jobs in factories and mines, places where people endured hardship for the promise of progress. Populations increased rapidly in areas where there was employment. The first industrial cities were born.

A major advance at this time was the discovery of coke, a sub-stance derived from coal. When burned, coke reaches extremely high temperatures, allowing increased production of iron. This greater availability of iron led to the construction of even more and larger machines.

During the 1700s industry continued to evolve. Steam engines — now improved even more by James Watt — were put to many new uses. By 1783 the first working paddle-wheel steamboat was chugging up a French waterway. Not long after, people started using "town gas"

IRON WORKS

Of all the advances that fueled the Industrial Revolution, one of the most important was an improvement in methods for producing iron. Heavy machinery made of iron played a key role in the growth of manufacturing, and coal was the energy source that made it possible.

The process of extracting iron (or other metals) from rock or ore is called *smelting*. Smelting iron from rock requires extremely high heat. The first furnace capable of generating this heat was the blast furnace, developed in Europe in the fourteenth century. It used charcoal (made from wood) as fuel. Modified over time, blast furnaces were burning hot enough to actually melt iron ore. The melted iron then separated from the ore and ran to the bottom of the furnace, where workers could collect and shape it (producing what we still know today as cast iron).

Coal was more widely available and burned hotter than charcoal. But the sulfur in regular coal made iron too brittle when smelted. In 1709 the development of coke — coal with the sulfur removed — allowed the use of coal in blast furnaces, revolutionizing the iron industry.

The demand for coal was greater than ever, as blast furnaces were fired up daily, churning out tons of iron (and later, steel, an alloy of iron and carbon) to make industry's increasingly complex machinery.

Iron ore, limestone, and coke

Blast furnace

Burning coke, ignited by blast of hot air

Hot air

Slag

Molten iron

Blast furnace

Industrial city

made from coal to light streetlamps in Cornwall, England. These were
both major breakthroughs. Now goods and people could travel faster,
and work could continue after dark.

By the early 1800s steam engines had become bigger and better
than ever, with five times the pressure of Watt's early engines. They
were now powerful enough to drive the mighty locomotives that made
history as they hurtled across continents in record time.

Coal-fired steam engines puffed away in factories, with smokestacks towering over most city landscapes by mid-century. Sources of
energy that had once seemed so dazzling and fantastic had become
not only commonplace, but absolutely essential.

A "STIRLING" IDEA

Robert Stirling of Scotland
designed an engine in 1816
that worked without burning a fuel.
It used heat (as from the sun) to
expand and contract air, causing a
piston to move up and down.
However, Stirling's invention was
overlooked and crowded out by the
already popular steam engine.
Practical uses for the Stirling engine
would not be developed until almost
200 years later. (See "Solar Dish
Engines," page 88.)

EXPERIMENTS WITH ELECTRICITY

Even before the end of the eighteenth century, electricity had entered the picture. In the mid-1700s Charles Dufay, a Frenchman, and Stephen Gray, an Englishman, had both conducted important experiments investigating electricity. And, in America, Benjamin Franklin's famous kite-flying demonstration had proven that lightning is electricity.

In the early 1800s Italian Alessandro Volta produced electricity from a wet-cell battery for the first time. During the same period American Joseph Henry, Englishman Michael Faraday, and Danish physicist Hans Øersted began experimenting with electromagnetism.

By the 1830s scientists had demonstrated that electricity and magnetism could be converted into one another. Soon huge electromagnets were lifting weights of more than a ton (2,000 pounds, or 907 kilograms), and early generators, called dynamos, were producing electricity by spinning magnets between wire coils. Electromagnetism was put to use in the telegraph to transmit messages tapped in Morse code. This electric communication — along wires and across long distances — was one of the first practical applications of electrical energy.

FOSSIL FUELS POWER INDUSTRY AND TRANSPORTATION

In the mid- and late- 1800s industry grew and factories spread. The Industrial Revolution was in full swing — especially in Great Britain, Germany, France, and the United States. The demand for coal and other fuel increased. To meet this demand, machines were developed to extract coal from the earth more quickly and more efficiently than ever before. In 1859 another type of fossil fuel became accessible, when the first oil well was drilled in the United States.

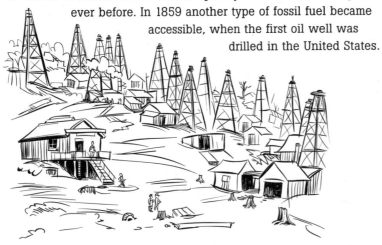

First oil derricks in the United States

THE FIRST GENERATOR

Michael Faraday was a 19th century British inventor with a fascination for electromagnetism. In 1831 he made important discoveries that led to his invention of the electric motor. In one of his most famous experiments, he took a horseshoe magnet and rigged a copper disc between the poles, positive and negative, then put the disc in contact with wires. By spinning the disc he was able to create a direct current in the wires. This was actually the very first generator.

Faraday was a very influential scientist whose work had lasting importance. In addition to his work on electromagnetism, Faraday was the first person to discover the laws of electrolysis. (See "By Electrolysis," page 108.)

The following year, in Belgium, gasoline (refined from oil) was used in the earliest working internal combustion engine, paving the way for development of the automobile.

In 1885 Karl Benz unveiled the first motorcar. Made in Germany, it was a three-wheeled, gasoline-powered model. A fellow German, Gottlieb Daimler, followed two years later with a four-wheeled version. Soon after, Frenchmen Edouard and André Michelin developed the first air-filled tires, which were easier to make and to use than the original solid rubber tires. Each year automobiles became more powerful and more popular. And with each passing year, demand increased for the gasoline needed to run them.

People wanted to travel faster and farther. Before long, coal became a fuel for steamships. In 1897 Englishman Charles Parsons installed a steam engine in his boat, *Turbinia*, and outran every ship in the water — even the huge three-masted clippers that, until then, had been the fastest ships at sea.

The *Turbinia*

ELECTRIC POWER - A CHANGING WORLD

This period of rapid growth in fossil fuel use coincided with the development of electric power. By early 1870 Belgian Zenobe Gramme had perfected the dynamo, making it the first practical electricity generator. A decade later, Parsons developed an even more efficient steam engine for powering these new generators. Widespread use of electricity — a so-called miracle of modern science — was now possible.

During this time Thomas Edison developed many handy uses for electricity. In 1879 he perfected the light bulb, and just a few years later Edison's first light and power operation, the Manhattan Pearl Street Station, opened in New York City. Edison's power plants produced direct current (DC), which could travel only very short distances. Serbian-American rival Nikola Tesla conceived of and developed a way to produce alternating current (AC), a process that would eventually allow electricity to travel far and wide. (See "Send in the Alternate," page 24.)

In 1882 the nation's first hydroelectric power plant was built in Appleton, Wisconsin, where it produced just enough electricity to light two paper mills and one home. In 1896 America's earliest large-scale hydropower plant was completed at Niagara Falls, New York. Since it produced AC, its voltage could be increased by transformers until it was high enough to transmit power many miles away to the city of Buffalo, without significant losses. The nation's first long distance transmission of electricity had occurred in California one year earlier, from Folsom to Sacramento. Before long over 200 electric power plants in the United States were producing all or part of their electricity using hydropower. Electric transmission and distribution lines soon became a regular feature of the American landscape.

Hydropower plant

The appetite for convenient electric power — often provided by coal-fired, steam-driven power plants — continued to grow. Large cities were putting electricity to another use — powering trolley lines for a convenient form of urban transportation. By the end of the nineteenth century there was widespread reliance upon electricity. Life had changed dramatically.

MEETING THE NEEDS OF THE TWENTIETH CENTURY

The 1900s brought even more changes. Cheap power and improved metals made mass production possible, and in 1908 the first inexpensive car — Henry Ford's Model T — rolled off the assembly line in the United States. In the same decade a pioneering power plant in Italy produced electricity using geothermal energy, and the first solar photovoltaic cells were invented.

During this time, Albert Einstein formulated his revolutionary theory about the relation of mass to energy. Einstein's concept of relativity opened the door for others to later discover a process by which mass can be converted to energy, paving the way for the future development of nuclear power.

SEND IN THE ALTERNATE

Electricity is transmitted in two ways: *direct* or *alternating*. With direct current (DC) the electrons flow in one direction. With alternating current (AC) the flow changes direction, oscillating back and forth in the conductor. In the United States, AC current does this 60 times (cycles) a second or 60 Hertz.

In the early days, DC was king; but DC current couldn't travel far along power lines. Then, in 1885, Americans George Westinghouse and William Stanley improved Nikola Tesla's electric alternator for producing AC current. AC could be transformed to a higher voltage, allowing it to travel anywhere transmission lines could go.

Just a few years into the twentieth century, hydroelectric power was producing 15 percent of America's electricity. In the 1920s, small wind turbines were also producing electricity, primarily in rural areas in the United States, France, and Denmark. But it was fossil fuel energy that dominated the power scene. Coal, oil, and natural gas were inexpensive and convenient, and people believed that these resources were plentiful. With fossil fuels readily available, the United States led the way in manufacturing, contributing 35 percent of all the industrial goods produced in the world.

By the 1930s automobiles swarmed over the countryside. The total mileage of surfaced roads soon exceeded that of railroads. And almost every home, office building, and factory had electricity.

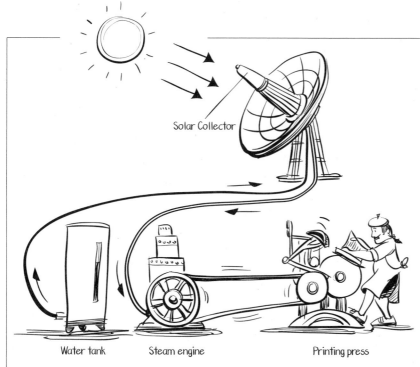

Solar Collector

Water tank   Steam engine   Printing press

A solar steam engine runs a printing press at the Paris Exhibition in 1878.

THE FIRST SOLAR ENGINE

In 1861 Auguste Mouchet patented the world's first solar steam engine. It used mirrors to focus heat from the sun onto a boiler to make steam for the engine. In 1878, at the Paris Exhibition, Mouchet and fellow inventor Abel Pifre used a solar steam engine to run a printing press. Pifre, a newspaper publisher, later demonstrated how efficient the solar engine printing press was by printing 500 copies per hour of his *Soleil* (Sun) *Journal*, in a large public garden in Paris. Unfortunately, engines like these were crowded out when the gasoline-powered engine was perfected. It would be decades before solar power would once again be used to create steam to do work.

NEW SOURCES OF POWER

The late 1930s and 1940s saw the first development of nuclear
technology. German Otto Hahn and Austrian Lise Meitner, among other
scientists, unveiled the process of nuclear fission — the release of
energy caused by splitting uranium atoms. Italian Enrico Fermi worked
in the United States to design and build the earliest nuclear fission
reactor. The world's first nuclear-powered electricity plant opened in
1954 in the former Soviet Union. Before long, dozens of nuclear power
plants were providing electricity for countries all around the world.

The last half of the twentieth century saw significant progress
in the development of wind, solar, and geothermal energy technologies.
These renewable resources all produce electricity without burning
fossil fuels. During the 1950s solar photovoltaic cells were improved
to generate electricity reliably, and NASA began using these and
hydrogen fuel cells in its space programs. The first American geothermal
power plant began operating in 1960, generating electricity from
natural steam brought up from wells drilled deep underground in
northern California.

Even though new energy technologies were developed, fossil
fuels still held the lead in industry, transportation, and generation of
electricity. By the early 1980s three out of every four power plants in
the United States burned fossil fuels. Almost every family had at least
one car or used public transportation fueled by gasoline or diesel.

WHERE DO WE GO FROM HERE?

Today, early in the twenty-first century, over 70 percent of the energy
we use in the United States comes from fossil fuels. There are long-
standing and growing concerns about our dependence on these
rapidly diminishing resources.

Fortunately, there are alternatives to the overuse of fossil fuels.
We can conserve energy, we can make our energy use more
efficient, and we can take advantage of renewable
energy resources. We can make energy
decisions that will improve the
environment, our lives, and
the lives of future
generations.

CHAPTER 2

ENERGY AND ELECTRICITY
How we produce and deliver most of our electricity

TERMS IN GLOSSARY

ampere (amp)
atom
baseload power
blackout (brownout)
Celsius
central station power plant
complete circuit
condenser
conductor
demand
electric current
electrical energy
electron
energy conversion (transformation)
energy storage
Fahrenheit
generator
grid
heat (thermal) energy
kilowatt, kilowatt-hour
magnetic field
mechanical energy
megawatt
negatively charged
neutron
nucleus
peaking power
positively charged
power load
proton
resistance
smart meters
static electricity
substation
transformer
transmission lines
turbine
vaporize
velocity
watt, watt-hour

HOW OFTEN DO YOU THINK ABOUT where your electricity comes from? If you are like most of us you don't give it too much thought, since most people in industrialized countries are many miles of wire removed from the places where their electricity is generated. (Those of us lucky enough to have our own way of generating electricity — such as from a wind turbine or solar panels — are still the exception.)

Behind the scenes, energy producers are working day and night to provide us with a steady supply of electrical power. Using improved technology and know-how, today's electricity suppliers have figured out many different, and sometimes complex, ways to generate electricity. But the most common and widespread method uses an age-old apparatus, the turbine, attached to a much more modern device, the generator. For over 120 years these two machines have worked together in power plants to produce vast quantities of electricity, revolutionizing the way people live, work and play.

A TYPICAL POWER PLANT

A turbine, a generator, and a source of energy make up a power plant, no matter how large or small. Even a single wind turbine can be thought of as a power plant. But usually when we think about electricity generation, we picture a huge steam-driven power plant filled with giant turbine generators, humming with force and energy.

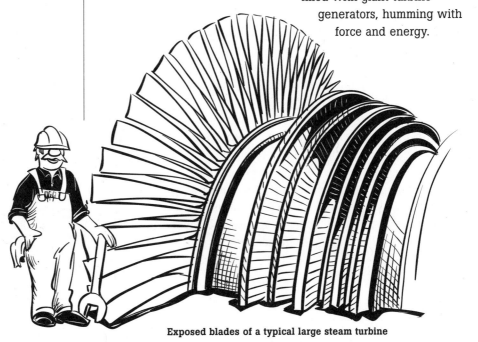

Exposed blades of a typical large steam turbine

The Turbine

A turbine is any device with blades attached to a central rod, or rotor, that spins when a force hits the blades. This spinning motion can do a lot of useful work. Waterwheels and windmills were actually our first turbines. Their wooden blades captured the power of wind or rushing rivers to lift water for irrigation or to rotate great stones to grind grain.

It wasn't until the 1880s, when the generator was first invented, that people began using turbines to produce electricity. Today we have many turbine designs. Some are small or have just a few main blades attached to the rotor (wind turbines, for example). Some turbines (such as those that use steam) are enormous, standing much taller than the average person. (See illustration, page 27.) These very sophisticated turbines have thousands of different-sized blades attached in a complicated pattern to the central rotor. These huge turbines are the kind used in most of today's large "central station" power plants.

Turning the Blades

The force of high-pressure steam powers most of today's turbines. We usually make the steam by burning a fuel such as coal, natural gas, or wood products to heat water above its boiling point.

But we don't always have to burn something to produce heat for making steam. The heat can come from the sun (in a solar thermal plant), from deep underground, using the earth's natural hot water and high-pressure steam (in a geothermal power plant), or from nuclear reactions (in a nuclear power plant). And forces other than steam, such as falling or running water, the wind, and ocean waves and tides, can also spin turbines. There are even ways we can generate electricity without using turbines at all. (See Chapter 3, "Energy Sources for Electricity Generation.")

Generating Electricity Using Electromagnetism

In a power plant, the sole function of a turbine is to spin a generator. A generator changes the mechanical energy of spinning to electrical energy. In a generator, coils of copper wire attached to the rotor spin inside a space surrounded by huge stationary magnets — or, as an alternate design, the magnet spins inside coils of wire. The magnetic field causes electrons in the wire to move, creating an electric current.

WHAT IS ENERGY?

Energy is the capacity to do work — to move something, heat it up, or change it in some way.

When work is being done, energy is always *converting*. For example, when you run, your body converts chemical energy from food you've eaten into the energy of your actions — mechanical energy — and heat. Often there are several energy conversions, in essence, an "energy chain."

A steam-driven power plant has a series of energy conversions in an energy chain that goes like this: Heat energy (to make steam from water) is converted to mechanical energy (the spinning of the turbine blades); and then it's converted again in the generator to electrical energy.

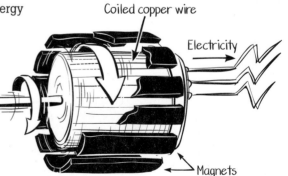

Coiled copper wire

Electricity

Magnets

A generator

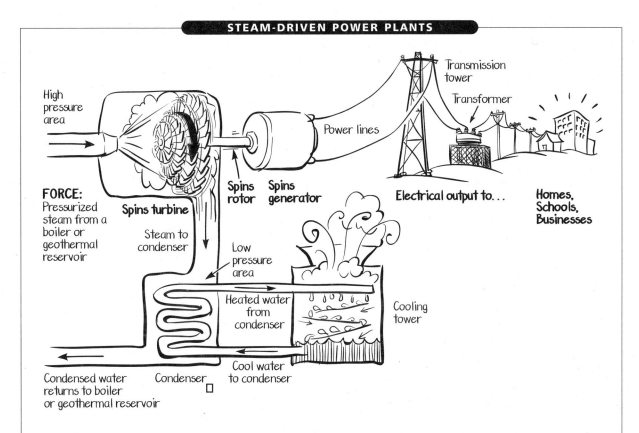

STEAM-DRIVEN POWER PLANTS

HOW A STEAM-DRIVEN POWER PLANT WORKS

The force of steam drives most of today's power plants. A jet of high-pressure steam spins the turbine, which, in turn, spins the generator — producing electricity. The electricity speeds through power lines that connect with our homes, schools, businesses and industry.

How does steam get its force? Water's boiling point is 212°F (Fahrenheit) or 100°C (Celsius). Below this temperature water is in a liquid phase, with the molecules packed densely together. But when water is heated above its boiling point, it "flashes" into steam (its gas, or vapor, phase), where the molecules separate and bounce around. Because the molecules are spread out, steam occupies a much larger space than it did as water — *over a thousand times as much space!* When steam is confined and not allowed to expand as it is vaporized, it becomes high-pressure (pressurized) steam.

In a turbine, high-pressure steam is released from a confined space. The steam bursts out through nozzles at very high velocity. It can reach speeds over 450 mph (miles per hour) — 724 km/h (kilometers per hour) — and blasts against the turbine blades. To get even more force, low pressure is created by cooling the steam as it rushes out of the turbine. This cooling condenses the steam back to water (contracting the space occupied). The push of the high-pressure steam and the pull of the low-pressure area create that extra oomph needed to spin the turbine even more efficiently.

UNDERSTANDING ELECTRICAL TERMS

Electrons traveling along a wire can be compared to water molecules flowing through a pipe. *Voltage* (expressed in volts, named after the scientist Alessandro Volta) is what pushes the electrons, like pressure pushes the water. Voltage is the force with which a source of electric current, such as a generator or battery, moves electrons. At power plants, electricity is usually generated at around 20,000 volts. By comparison, the light bulb in your desk lamp operates at 120 volts.

Current is the rate of flow of electric charge. It is the number of electrons flowing past a given point per unit of time (usually one second). The amount of current flowing in a wire is expressed as *amperes*, or amps (named after Andre Marie Ampere). This is similar to describing water flow in gallons per minute. An ampere is equal to about 6.25×10^{18} electrons per second. That's a LOT of electrons — 625 followed by 16 zeros.

A *watt* (W) is a unit of power (named after James Watt). It is the rate at which work is performed. One watt is the rate of current flow when one ampere is "pushed" by one volt. One watt is needed by a typical string of Christmas lights. A kilowatt (kW) is 1,000 watts, the average amount used by homes in the U.S. One megawatt (MW) is 1,000 kilowatts (1 million watts). In the U.S., 1 MW serves approximately 1000 homes.

The electricity industry also uses the terms *watt-hour* (Wh) and *kilowatt-hour* (kWh) to measure electricity use. A watt-hour is the amount of electricity used in one hour by a device that requires one watt of power to operate. A kilowatt-hour is 1,000 watt hours. For example, a 100 watt light bulb that is left on for one hour will use 100 watt-hours of electricity. If left on for ten hours, the same bulb will use 1,000 watt hours, or one kilowatt-hour.

WHAT IS ELECTRICITY?

Atoms — the tiny building blocks of all matter — are electrically neutral, or uncharged. The center, or nucleus, of an atom has neutrons and positively charged protons. It is surrounded by an equal number of negatively charged electrons. These charges balance each other to make the atom electrically neutral.

Certain kinds of atoms have electrons that are loosely held. The electrons can be attracted to move from one atom to another. The movement of these electrons over time causes an electric current (electricity). (Scientists still aren't certain what this "movement" is.)

Electricity needs a pathway to follow, called a complete circuit. The electrons flow from negative to positive along this pathway. Some types of metal wire allow electrons to flow from atom to atom more freely than other types. These are good electricity conductors because they have less resistance to the flow of electrons. Copper wire has low resistance and so it is commonly used for electrical wiring.

Electricity is known as an *energy carrier*; the electrons carry energy in a useable form from one place to another.

SENDING ELECTRICITY TO CUSTOMERS

Electricity is sent, or transmitted, from power plants along power lines. Power lines include high-voltage transmission lines and a complex system of smaller distribution lines serving communities.

The Power Grid

The entire interconnected system that distributes electricity — power plants, transmission and distribution lines, towers, substations, and transformers — is called the power grid. Most power grids cover large regions, sometimes encompassing several states. Grids operated by neighboring utilities are generally connected to each other for a smooth flow of electricity from region to region.

Increasing and Decreasing Voltage

Before electricity leaves the power plant, a transformer increases, or steps up, its voltage. Increased voltage improves efficiency over long distances. For safety, high-voltage transmission lines are installed on tall towers or underground. You may have been close enough to one of these to hear the crackle of high voltage electricity in the wires.

As the electricity nears its destination, it goes through a substation where the voltage is lowered, or stepped down for distribution. But the voltage is still too high to be used directly in your home or business. So, before entering a building, the current goes through yet another smaller transformer to drop the voltage again. These small transformers are often mounted high on a utility pole near the building. Some of us can go outside and see these small transformers, which, in the U.S., normally change the voltage from several thousand volts to 120 and 240 volts. (Some appliances need higher voltage than others.)

The lines also pass through a watt-hour meter, usually located on the outside of the building. This meter measures and displays electricity use in kilowatt-hours.

ENERGY STORAGE

Some electricity comes from energy sources that are "part-time" forces — like the sun and the wind. (See Chapter 3.) We can't change when the sun shines or when the wind blows, but to some extent we can store the electricity they generate. By using large banks of big batteries (like car batteries), we can store power from the sun (solar power) to use at night or from wind power to use on a calm day. Much research is focused on improving batteries to store electrical power.

Batteries are not the only way to store energy. In "pumped hydro storage," water is pumped uphill to high reservoirs during off-peak hours; then, during peak periods of power demand, the water is released to drive hydropower turbines. (See sidebar, page 64.) Other energy storage systems include flywheels and compressed air systems.

MANAGING THE LOAD

Our electricity is billed in kilowatt-hours. Most meters track the amount of electricity we use, but not the times that we used it. New "smart meters" will allow us to see and track our electricity use in real-time. The time of day that we use electricity is important — for the electricity suppliers, the grid managers, and for the electricity customers.

There are many strategies that can be implemented to manage the electricity load. Chapter 5 of this book has information about these strategies, including alternatives for the transmission and distribution of electricity.

Baseload Power

The amount of electricity generated depends on the demand (how much is being used) at any one time. The basic amount of electricity that must be produced all the time (the amount that the grid managers know will be needed night and day) is called *baseload power*. Supplies of baseload electricity can be planned far ahead. Baseload power is traditionally supplied under long-term contracts from large power plants that operate 24 hours a day. It is traditionally the least expensive electricity.

Peaking Power

The electricity demand above the baseload is called *peaking power*. The need for peaking power can fluctuate greatly, depending on the time of day, week, or season. The highest loads are often on cold winter days when we need more electric heat, and on hot summer afternoons when we need the most cool air conditioning.

Peaking power is more expensive than baseload power. This is because peaking power plants are turned on only when they're needed (so they can sell electricity only part-time). And sometimes peaking power has to be purchased at high cost from out of the region.

Problems can occur when our demand for electricity exceeds the amount of peaking power available. If you have experienced electrical "blackouts" or "brownouts" it might have been because there wasn't enough peaking power available to cope with a sudden increase in demand where you live.

ELECTRICITY CHOICES

In the following chapters you will learn about different ways we can meet our electricity needs. We have choices to make — about how much electricity we use, when we use it, and the energy resources we use to generate it. These choices are becoming more important every day.

CHAPTER 3

ENERGY SOURCES FOR ELECTRICITY GENERATION
How we use different energy sources to produce electricity

TERMS IN GLOSSARY

alternative energy

biomass

capacity

deplete

fossil fuels

geothermal energy

green energy

hydrogen gas

hydropower

nonrenewable energy

nuclear fuels

ocean energy

regenerate

renewable energy

solar energy

sustainable

wind energy

IT'S EASY TO TAKE our seemingly plentiful supply of electricity for granted, especially in the United States. We can flick on our lights or get a cold drink from our refrigerators just about anytime we want. Since we seem to have so much electricity, we might conclude that the energy sources we use to generate this electricity are also found in abundant quantities; but this is only partially true. Renewable energy sources will always be available, but others — the nonrenewables — are being used up.

RENEWABLE AND NONRENEWABLE ENERGY SOURCES

Renewable energy sources are those that are naturally regenerated, or renewed, within a useful amount of time: wood and other substances produced by living things (biomass), natural heat from the earth's interior (geothermal), streams and rivers (hydropower), the wind, the sun (solar), and the ocean. We can use these resources today, and nature will still provide them tomorrow.

Nonrenewable energy sources are those that can be depleted. Nature does not renew these in a useful amount of time. These include fossil fuels (coal, oil, natural gas) and nuclear fuels. These resources are being used faster than nature could ever replace them.

Renewable energy sources

Energy Resources for Electricity Generation

Renewable Energy Resources

 Biomass: Plant material (including wood) or organic waste

Geothermal: The natural heat in the earth

 Hydropower: The force of moving water from streams, rivers or storage reservoirs

Ocean: The mechanical energy of ocean tides, currents, and waves, and the sun's heat energy stored in the ocean

 Solar: The radiant energy from the sun

Wind: The force of moving air

The Renewable and Nonrenewable Resource

 Hydrogen: Hydrogen gas produced from other energy resources; an energy carrier

Nonrenewable Energy Resources

 Fossil Fuels: Coal, oil (petroleum), and natural gas

Nuclear Fuels: Elements with unstable nuclei, such as uranium

RENEWABLE? CLEAN? GREEN?

We sometimes read or hear the terms "clean energy," "green energy," "sustainable energy," and "alternative energy," along with the term "renewable energy." Some people use these terms interchangeably, which can be confusing.

Clean or *green* energy usually refers to energy that is environmentally friendly. When we generate electricity with these resources, very few pollutants, if any, enter our air or water.

Sustainable energy usually refers to a process, system, or technology that does not deplete resources or cause environmental damage. Sustainable energy practices preserve meaningful natural resource choices for future generations.

When people use the term *alternative* energy, they are usually speaking of alternatives to the conventional energy sources, which are fossil fuels, "large" hydropower, and nuclear. Alternative energy definitely includes renewables. More often, though, the term "alternative" is applied to certain transportation fuels — those other than gasoline and diesel, such as ethanol, biodiesel, and hydrogen.

ENERGY AWARENESS, ENERGY CHOICES

Electricity has contributed greatly to our comfort and to our society's development, but we are using up valuable and finite energy resources. Since the beginning of the Industrial Revolution our use of energy sources, particularly fossil fuels, has increased with each passing year. In the last 30 years alone, their use has tripled. Some worldwide experts believe that our ability to produce oil has peaked and that current rates of production cannot be sustained.

We are indeed fortunate to have other energy options. In the pages that follow you will find a comprehensive explanation of the energy resources and technologies we use to make electricity.

SIZING IT UP

In this chapter, power plant sizes (in kilowatts and megawatts) are given for each energy resource. A power plant's size is the amount of electricity it can produce at any one time. This is known as a plant's "capacity." But power plants do not always operate at full capacity. The amount of electricity actually produced over time depends on many factors. Some of these factors are addressed in the "Considerations" at the end of each resource section.

The percentages given in the pie charts on the next page are from actual electricity produced.

RESOURCES BEING USED TO GENERATE ELECTRICITY
These charts show the percentages of electricity produced from
different energy resources in the United States and around the world.

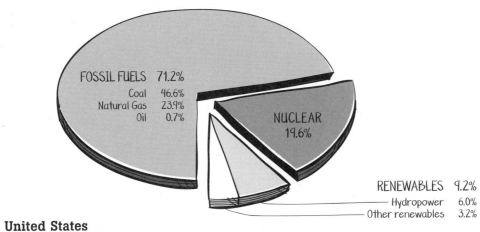

FOSSIL FUELS 71.2%
Coal   46.6%
Natural Gas 23.9%
Oil   0.7%

NUCLEAR
19.6%

RENEWABLES 9.2%
Hydropower   6.0%
Other renewables 3.2%

United States
Year 2008

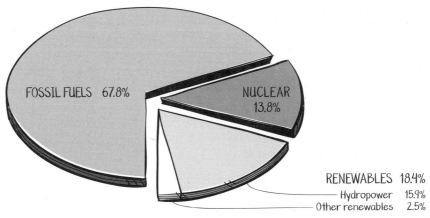

FOSSIL FUELS 67.8%

NUCLEAR
13.8%

RENEWABLES 18.4%
Hydropower   15.9%
Other renewables 2.5%

World
Year 2007

Source: U.S. Energy Information Administration, 2010

Renewable Energy Sources

Renewable Energy Source:
BIOMASS

TERMS IN GLOSSARY

anaerobic digester

by-product

carbon cycle

cogeneration

decompose

energy farm

gasification

green waste

methane gas

microbe

soil erosion

BIOMASS WAS ONE OF THE FIRST energy resources ever used by humans. It includes anything that is or was once alive. Since the discovery of ways to create fire, humans have been burning wood and other organic material to create heat and light.

In the United States, biomass, mostly from trees, was the premier energy source until the 1830s. It was displaced by fossil fuels (mainly coal) when the Industrial Revolution took hold. Recently, however, the use of biomass, in a widening range of forms, has begun to increase. Today it is an important energy source for many processes, including the generation of electricity.*

THE BIOMASS RESOURCE

Most living things receive and store energy from the sun. This energy is released when the organic material is digested, burned, or decomposed. This released energy can be used to produce electricity. Today, many kinds of biomass are used as energy resources.

Solid Biomass

Solid biomass is anything organic that has not yet broken down into a gas or a liquid. There are many kinds of solid biomass. Chipped wood, whole trees, energy crops, and agricultural wastes are examples. Other solid biomass sources are trimmings from forests and orchards; wastes from building construction, food processing, and paper industries; animal manure; and plain old garbage.

At home, and at work, people produce tons of waste each year, much of which is organic. Many of us produce a lot of this "green" waste just from cutting our lawns and trimming our trees and bushes.

Until relatively recently, all garbage (including organic waste) was dumped into landfills or burned without any pollution controls. Today, many biomass power plants (complete with pollution controls) use solid biomass to produce electricity. Instead of going to landfills much of our green waste is now trucked directly to biomass plants. A plant in Michigan, for example, uses 300,000 tons per year of wood waste from local timber industries (and puts wastewater to use in its cooling towers). A plant in Wisconsin uses 250,000 tons of wood wastes, shredded railroad ties, and even scrap tires to produce electricity.

Fast-growing trees are ready to be harvested for use in a biomass power plant.

Biomass is also used for space heating and factory processing and to produce liquid transportation fuels such as ethanol.

POWER SKETCH: Munching Microbes

Picture a landfill teeming with rotting, long-buried waste. Microbes gobble this decaying quagmire of leftover stuff that originally came from living things. As the microbes munch, they burp methane gas. Methane gas is normally released into the atmosphere and is a potent greenhouse gas (see Glossary). However, at a landfill near Eugene, Oregon, as at many others around the United States, the gas is collected and burned for heat to generate electricity. This biomass power plant has been in operation since 1992 and continues to send electrical power to several thousand homes.

Biofuels and Biogas

We can produce both liquid and gas fuels from solid biomass. This is not a new idea. The production of biomass gas, called gasification, is based on a method developed in the early 1800s to produce gas from coal for town streetlights in both England and the U.S. And since the 1940s, in over a million homes in India, people have cooked with biomass gas made in their own small gasifiers.

Today, gasifiers use high-tech processes to produce a gas from solid biomass by heating it under very controlled conditions. This gas can then be converted to a liquid. Gasification facilities can be large or small, serving power plants that range from just a few kilowatts to 50 MW or more.

WASTE TO ENERGY

A biomass power plant in Shasta County, California, processes about 90 tons of solid waste from timber mills, forests, and orchards every hour, producing enough electricity to power 50,000 homes. Each day at a biomass power plant in Vermont, about 200 tons of waste wood from local forests are converted to gas, which is burned to produce "homegrown" electricity.

Some biomass gas occurs naturally. Leftover biomass will decompose on its own, producing gases such as methane (a colorless, flammable gas). These gases can be collected for use in a biomass power plant. Some of these plants are located at landfills to burn the gas right as it's formed. Of the estimated 2,500 municipal solid waste landfills throughout the United States, over 400 are now the sites of landfill gas power plants, with plans for dozens more.

Near Vancouver, British Columbia, a new landfill gas cogeneration power plant produces electricity and, at the same time, supplies heat to a greenhouse for growing tomatoes. Denmark has solved its livestock manure problem by turning most of it into a biomass gas fuel for heating and for generation of electricity.

> **REMINDER**
>
> **W** = watt
> **kW** = kilowatt = 1,000 watts
> **MW** = megawatt = 1,000 kilowatts
> 1 megawatt can serve about 1,000 homes in the United States.

Biomass Energy Farms

Sometimes specific crops and trees are grown just for biomass power. These are often referred to as biomass energy farms. Hybrid willow and poplar trees as well as switchgrass are the crops most widely used today. They grow fast, help keep loose soil from eroding, and thrive in a variety of growing conditions. Hybrid willows and poplars can be cut and used for energy as often as every three years, as they regrow quickly from the cut stumps.

For many years, farmers have been growing switchgrass as a side crop for livestock feed and to control soil erosion. Now, some of these farmers are growing switchgrass as their main crop — an energy crop. For example, in Alabama, farmers are successfully raising switchgrass energy crops in soil once depleted and eroded by the over-harvesting of cotton.

Besides providing a local, abundant, and green energy source, growing energy crops can also revitalize the economies of rural areas. It has been estimated that the United States has sufficient available land to grow enough biomass to supply one fourth of its current energy needs.

Switchgrass, a biomass energy crop, swiftly grows to 10 feet high.

ELECTRICITY FROM OUR TRASH

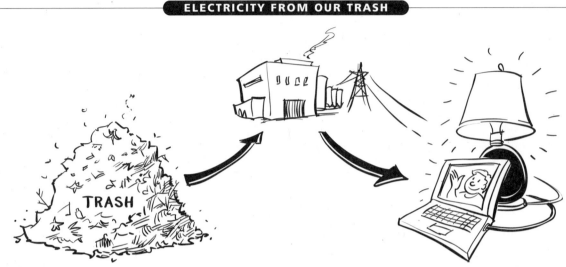

What most of us call "garbage," energy professionals call "municipal solid waste." Today, some companies are using this waste (which otherwise would be wasted!) as a fuel source to bring us heat and electricity.

Our garbage contains biomass materials like paper, cardboard, food scraps, grass clippings, leaves, wood, and leather products. (Even though many of us are diligent about recycling most of these materials, there are still millions of people who, unfortunately, still throw everything in the trash.) This garbage also contains combustible materials that are not traditionally considered "biomass" — mainly plastics and other synthetic materials made from petroleum.

Today, we use about 7 percent of the country's waste in special "waste-to-energy" power plants. There are close to 90 of these waste-to-energy plants in the U.S., generating enough electricity to supply over a million and a half households — and disposing of the waste created by 20 million people!

Generating electricity at waste-to-energy plants has a big environmental bonus. Plants like these help reduce the amount of garbage we bury in our increasingly crowded landfills. And, as with all renewable energy sources, they help reduce greenhouse gases and provide green jobs to local communities.

For every ton of waste processed in a waste-to-energy power plant, we avoid the need to import one barrel of oil or mine one-quarter ton of coal.

GENERATING ELECTRICITY FROM BIOMASS RESOURCES

Biomass Power Plants

Biomass power plants usually work by burning organic matter or a biofuel to produce heat to boil water for steam to drive a turbine generator. These power plants vary in size.

Large Biomass Power Plants. Large-scale biomass power plants often resemble traditional steam-driven plants, such as those that run on fossil fuels. In a biomass plant, however, the energy production process includes the preparation and processing of the biomass for burning. If it's wood, it might be chipped. If it is garbage, non-burnable materials are removed, and sometimes the remainder is formed into pellets. At other biomass plants, the biomass is converted into a gas or liquid fuel before it is burned.

The processed biomass is then burned in enormous furnaces. The resulting heat boils water for steam that is used to drive turbine generators. Biomass power plants have special technologies that clean most of the ash byproducts and smoke produced from burning before they are released into the atmosphere. Like most other power plants, they have condensers that cool and condense the steam back to water. Then the water is cycled through the plant again.

IT'S A GAS!

While biofuels are often burned to heat water for steam-driven electrical generation, they can also produce electricity without creating steam. Biofuel gases themselves are sometimes used to drive gas turbines. These gas turbines are driven by the expansion of air and gases created by burning them in a confined space, rather than by using the heat to create steam. (See "Gas Turbines," page 121.)

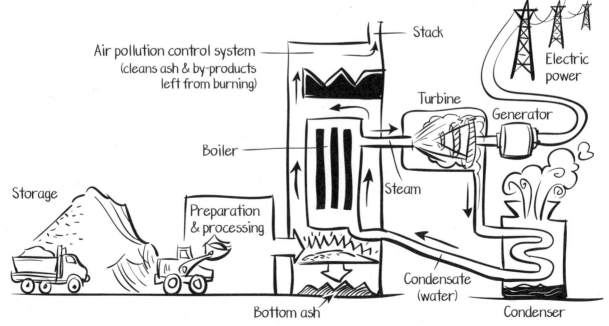

Large-scale biomass power plant

In Pietarsaari, Finland, one of the world's largest "biofueled" power plants produces up to 260 MW of electricity from forest residues. This plant produces steam for forest industries, as well as electricity for the local grid. Finland is actively promoting the increased use of biomass, as are many other European countries.

Small Biomass Power Plants. Small biomass power plants are often found in rural areas and in the villages of developing countries. These little powerhouses usually make use of locally generated biomass. They can deliver electricity to a single facility or to a small number of nearby users. Throughout the U.S., it is becoming more and more common for a dairy farmer to use cow manure to produce methane gas in an anaerobic digester (sometimes called a methane digester). The methane gas drives a biomass power plant that generates enough electricity for the dairy farm and — in some cases — even for neighbors.

The University of California at Davis is building a large anaerobic digester that combines manure from the school's animal science program with the school's garbage and green waste. The system is intended to satisfy the heat and electricity needs of 500 homes.

Cofiring. Biomass can also be burned along with another type of fuel, such as coal, in a process called cofiring. This can be done using existing equipment in a traditional coal power plant. The substitution of biomass for fossil fuels at these power plants reduces the amount of pollutants produced. Most of the electricity in the United States is produced from coal, so adding biomass to the mix can have a positive effect. Some coal power plants even dedicate a portion of their operations to burning only biomass.

FUELING AROUND IN SPAIN

Spain, the world's largest producer of olive oil, is the first country to produce electricity from olives. Since 1995, a power plant located amidst the olive groves of Andalucia has been producing enough electricity for 27,000 households. Growers in the area routinely turn their waste from olive oil production into biomass for power production.

CONSIDERATIONS

- Biomass energy crops are beneficial to the environment because they take in carbon dioxide (CO_2) as they grow. This can offset CO_2 — a greenhouse gas — given off when they are burned.
- Use of orchard and forest trimmings, along with other green waste for biomass fuel, can reduce waste disposal and landfill costs.

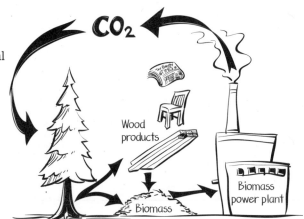

CO_2 from biomass power plants is offset by growing trees and crops.

- Anything that is burned gives off some byproducts (ash and gases, for example). Modern power plants that use solid biomass have special equipment that is effective in preventing pollutants from going into the atmosphere.

- Gases produced from decomposing organic material in landfills are pollutants and, if highly concentrated, are toxic. Collecting and burning these methane and related gases as fuel helps solve this problem.

- When transported or stored for use as a combustible material, solid biomass can take up a lot of space.

- Some people think that thinning overgrown forests and collecting fallen branches and tree trunks from the forest floor for biomass fuel protects forests from catastrophic wildfires and contributes to a healthier forest ecosystem. Others fear that, if not done correctly, this practice can adversely affect animal habitats and/or disrupt fragile ecosystems.

- Burning plastics and petroleum products results in emissions of dioxin — a toxin. In the past, such emissions during electricity production in waste-to-energy plants were cause for concern. But federal regulations, which went into effect in 1990, spurred the development of significant new technologies that combust waste at higher temperatures. These technologies have resolved the issues and have led to a 99 percent reduction in dioxin emissions.

- It actually costs more to generate electricity at a waste-to-energy plant than it does at other, more conventional, power plants, but most people think the environmental benefits outweigh the cost differences.

(continued)

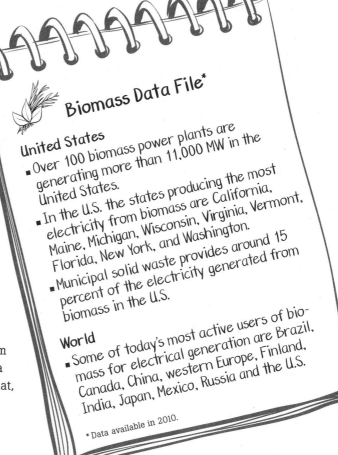

Biomass Data File*

United States
- Over 100 biomass power plants are generating more than 11,000 MW in the United States.
- In the U.S. the states producing the most electricity from biomass are California, Maine, Michigan, Wisconsin, Virginia, Vermont, Florida, New York, and Washington.
- Municipal solid waste provides around 15 percent of the electricity generated from biomass in the U.S.

World
- Some of today's most active users of biomass for electrical generation are Brazil, Canada, China, western Europe, Finland, India, Japan, Mexico, Russia and the U.S.

*Data available in 2010.

CONSIDERATIONS (continued)

■ Biomass is a renewable resource if we don't harvest the organic materials faster than crops or forests can be cultivated or naturally regenerated.

■ Biomass power plants are often used to supply baseload power because they can run day and night, and their energy supply is predictable.

Creating Electricity from Biomass Resources

Methane from Waste
This spotless room is in Exelon Power's Fairless Hills Generation Station, which produces electricity using methane from a local landfill. This power plant, which includes a Renewable Energy Education Center, received the Pennsylvania Governor's Award for Environmental Excellence. *(Photo courtesy of Noria Corporation, Tulsa, OK)*

Forest Waste
At Guadalajara's Alto Tajo Natural Park, forest waste from the park fuels this 2 MW generating plant. Besides its use for electric generation, the plant is to serve as a facility for research in biomass. *(Photo courtesy of Iberdrola Renovables, Spain)*

Biomass from Agricultural and Forest Waste
The fertile San Joaquin Valley, CA, produces great quantities of agricultural waste, and the adjacent Sierras produce forest waste. Old practices of open burning in the farmlands, and leaving the forests for wildfires, are being replaced gradually by biomass generating plants. Covanta Mendota, pictured above, processes more than 600 tons of waste each day and generates up to 25 MW hours of electricity. *(Photo courtesy of Covanta Energy Corporation, NJ)*

Creating Electricity from Biomass Resources

Chicken and Cow Power

Waste from farm animals pollute water and make a mess — unless they are used for power. Holland claims the world's largest plant running exclusively on methane produced from poultry manure. It uses a third of the nation's chicken waste to generate 36.5 MW of electricity. Many dairy farmers across the U.S. are using cow manure to produce methane gas to drive small biomass power plants, generating electricity for their farms. *(Chicken photo courtesy of Lena Szymoniak, Aspenlund Farm, MT)*

Switchgrass

Switchgrass grows quickly to heights up to 10 feet and one planting can be harvested for 10 to 20 years. Although intended to produce ethanol for transportation fuel, switchgrass is also being harvested to generate electricity. This photo shows switchgrass being harvested and baled. *(Photo courtesy of the University of Wisconsin-Madison)*

Cofiring with Biomass

Coal and woodchips combine with chipped tires (among other things) to provide fuel for a cofiring power plant at the University of Missouri. The plant, which provides electricity for the campus, uses over 7,000 tons of waste wood chips annually. The addition of this biomass to the mix reduces the amount of pollutants produced. *(Photo courtesy of the University of Missouri)*

Renewable Energy Source:
GEOTHERMAL

TERMS IN GLOSSARY

binary power plant
conduction
crust
dry steam power plant
enhanced geothermal systems (EGS)
fissure
flash power plant
fumarole
geothermal reservoir
groundwater
heat exchanger
hot dry rock
hydrogen sulfide
magma
mantle
modular
mud pot
porous
subducting
tectonic plates
wastewater
working fluid

PEOPLE HAVE ALWAYS BEEN FASCINATED with volcanoes and their fiery displays of nature's power. Many ancient societies believed volcanoes were homes to temperamental gods or goddesses. Today science tells us that volcanoes result from the immense heat energy — geothermal energy — found in Earth's interior. This heat also causes hot springs, fumaroles (steam vents) and geysers.

Over the ages, humans have benefited from Earth's geothermal energy by using the hot water that naturally rises up to the earth's surface. We have soaked in hot springs for healing and relaxation and have even used them as instant cooking pots. Hot springs have also been an important part of cultural life and healthy lifestyles, especially in Japan and Europe.

Today we drill wells deep underground to bring hot water to the surface. We use the heat energy from this geothermal water to warm buildings, to speed the growth of plants and fish, and to dry lumber, fruits and vegetables. (See "Direct Use Geothermal," page 149.) We use the energy from the hottest water to generate electricity.

POWER SKETCH: Fine Neighbor

Set amidst the open vistas and forests of the eastern Sierra Nevada of California, a power plant churns out enough electricity for about 40,000 homes. The natural setting is not marred by smoky emissions, because there are none. This geothermal power plant uses hot water resources from an underground geothermal reservoir to power its turbine generators. Many tourists and residents of nearby Mammoth Lakes don't even notice that the power plant is right there beside the main highway.

Mammoth Lakes geothermal power plant

THE GEOTHERMAL RESOURCE

Geo means earth and *thermal* means heat. Geothermal energy is the earth's heat energy.

The Inner Earth: Hot, Hot, Hot!

Billions of years ago our planet was a fiery ball of liquid and gas. As the earth cooled, an outer rocky crust formed over the hot interior, which remains hot to this day. This relatively thin crust "floats" on a massive underlying layer of very hot rock called the mantle. Some of the mantle rock is actually melted, or molten, forming magma.

The heat from the mantle continuously transfers up into the crust. Heat is also being generated in the crust by the natural decay, or breakdown, of radioactive elements found in most rocks.

The crust is broken into enormous slabs — tectonic plates — that are actually moving very slowly (about the rate your fingernails grow) over the mantle, separating from, crushing into, or sliding (subducting) under one another. The edges of these huge plates are often restless with volcanic and earthquake activity. (See "Hot Locations," page 56.) At these plate boundaries, and in other places where the crust is thinned or fractured, magma is closer to the surface. Sometimes magma emerges above ground — where we know it as lava. But most of it stays below ground where, over time, it creates large regions of very hot rock.

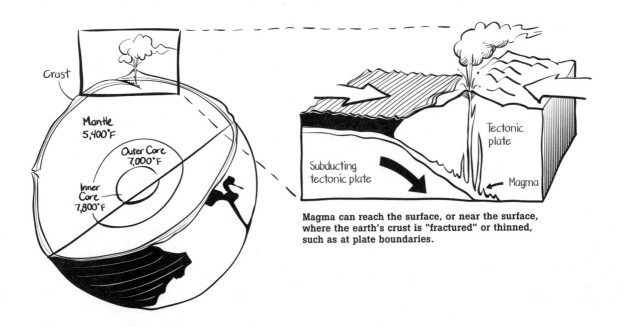

Magma can reach the surface, or near the surface, where the earth's crust is "fractured" or thinned, such as at plate boundaries.

Fumarole   Fumarole

Fault

Hot
Spring

WATER
SOURCE:
Deep
circulation of
rainwater
and
snowmelt

Fault

Cold water

Hot water

Heat

HOT

Hot Rock

GEOTHERMAL RESERVOIR:
Hot water in hot rock

Hot Rock

Very Hot Rocks

HEAT SOURCE:
Transfer of heat (thermal) energy from
cooling magma to surrounding rock and water

**A geothermal reservoir is a large underground area of
hot permeable rock saturated with extremely hot water.**

Geothermal Reservoirs: Earth's Natural Boilers

Rainwater and melted snow can seep miles below the surface, filling
the pores and cracks of hot underground rock. This water can get really
hot. It can reach temperatures of 500°F (260°C) or higher — well
above the normal boiling point of 212°F (100°C).

Sometimes this hot water will work its way back up (hot water
is less dense than cold and so tends to rise). If it reaches the surface
it forms hot springs, fumaroles, mud pots, or geysers. If it gets trapped
deep below the surface, it forms a "geothermal reservoir" of hot water
and steam. A geothermal reservoir is an underground area of cracked
and porous (permeable) hot rock saturated with hot water. The water
and steam from these super hot reservoirs are the geothermal resources
we use to generate electricity.

GENERATING ELECTRICITY FROM GEOTHERMAL RESOURCES

Geothermal reservoirs can be found from a few hundred feet deep to two miles or more below Earth's surface. To reach them, we drill wells and then insert steel pipe (casing). Now with an open passageway to the surface, the hot geothermal water or steam shoots up the well naturally or is pumped to the surface. From here it's piped into a geothermal power plant.

Geothermal Power Plants

There are different kinds of geothermal power plants, because there are different kinds of geothermal reservoirs.

Flash Steam Power Plants. Flash steam plants use really hot geothermal reservoirs of about 350°F (177°C) or higher. From the well, high-pressure hot water rushes through pipes into a "separator," where the pressure is reduced. This causes some of the water to vaporize vigorously ("flash") to steam, the force that drives the turbine-generators. After the steam does its work, it is condensed back into water and piped back down into the geothermal reservoir so it can be reheated and reused. Most geothermal power plants in the world today are flash plants.

Hot water or steam from a geothermal reservoir shoots up a geothermal well, spins the turbine-generator, and is returned to the reservoir.

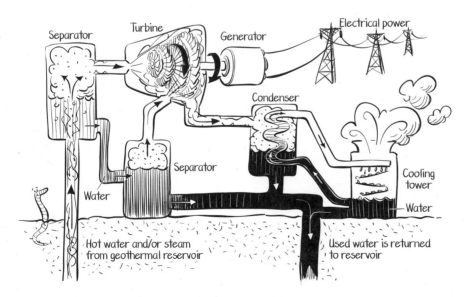

Flash steam plants can include one steam/water separator or, more commonly, two separators (shown here).

Dry Steam Power Plants. Very few geothermal reservoirs are filled naturally with steam, not water. This means that the wells will produce only steam. The power plants that run on this steam are called "dry steam" power plants. Here, the steam blasts right into the turbine blades (they do not need separators), then is condensed to water and piped back into the reservoir. Though dry steam reservoirs are rare, they have been important to the development of geothermal power, especially in California, Italy, and Japan.

Binary Power Plants. In some geothermal reservoirs, the water is hot (usually over 200°F, or 93°C), but not hot enough to produce steam with the force needed to turn a turbine-generator efficiently. Fortunately, we can generate electricity from these "moderate temperature" reservoirs using binary power plants. Moderate-temperature reservoirs are more common than high-temperature reservoirs, so the use of binary power plants is expanding worldwide.

In the binary process, the geothermal water is used only to heat a second liquid. After passing through a heat exchanger, the geothermal water is pumped right back into the reservoir. It is that second "working fluid" that flashes to vapor and drives the turbine. (See sidebar.)

HEAT EXCHANGERS

Heat exchangers are used in electricity generation when the heat source is hot, but not quite hot enough to bring water to a boil to create forceful steam.

A heat exchanger transfers heat (thermal energy) from a hotter liquid to a cooler one by conduction. The heat is conducted from the first (hotter) liquid into the second liquid (the "working fluid") through metal pipes or plates that keep the two liquids separated.

The second liquid is usually one with a lower boiling point than water, so it vaporizes, or "flashes" to vapor, at a lower temperature than does water. Sometimes, such as in certain solar thermal power plants, where the first liquid is oil or another material, the second liquid can be water.

The force of the rapidly expanding vapor or steam spins the turbine blades that drive a generator. The vapor or steam can then be condensed back to a liquid and used over and over again.

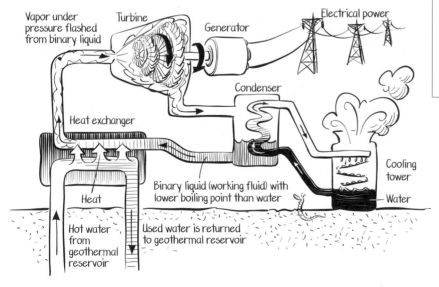

Binary power plant

All Shapes and Sizes

Geothermal power plants come small (200 kW to 10 MW), medium (10 MW to 50 MW), and large (50 MW to 100 MW and larger). A geothermal power plant usually consists of two or more turbine-generator "modules" in one plant. Extra modules can be added as more power is needed.

Binary plants are especially versatile because they can use relatively low reservoir temperatures. Small binary modules can be built quickly and transported easily. These little power plants are great for use in remote parts of the world, far from transmission lines. One interesting plant is installed in the rugged mountains of Tibet (People's Republic of China). At a soaring 14,850 feet (4,526 meters), it is the highest geothermal power plant in the world.

Small binary plants are also popular in sometimes remote hot spring spas and health resorts. They add the convenience of electricity while maintaining an environmental and healthful appeal.

REMINDER

W = watt
kW = kilowatt = 1,000 watts
MW = megawatt = 1,000 kilowatts

1 megawatt can serve about 1,000 homes in the United States.

HOT TIMES IN ALASKA!

Chena Hot Springs Resort has found an answer to high energy demands by tapping into the local geothermal resource. Chena is just east of Fairbanks — only 150 miles south of the Arctic Circle. Winters are extremely cold, with temperatures dropping to -50°F and colder. This means a LOT of energy is required for heating as well as for electricity. Chena is using a small binary power plant to generate 400kW of geothermal power with a resource that's only 165°F — the lowest geothermal temperature ever used for making electricity. The resort is also using geothermal water to make ice, heat buildings and greenhouses, and — of course — for the natural hot springs pools that make Chena famous.

Geothermal at Work Around the Globe

So far, the U.S. produces more electricity from geothermal energy than does any other country. Six states now have geothermal power plants. California has the most, followed by Nevada, Utah, Hawaii, Idaho, Alaska, Oregon, Wyoming, and New Mexico. More are planned in these states and are also being considered in Arizona, Colorado, Florida, Louisiana, Mississippi, and Washington.

U.S. geothermal power plant types vary widely. One little 300 kW geothermal powerhouse in northern California runs all by itself and automatically radios an operator when it needs maintenance. At one of Nevada's geothermal plants, the heat from geothermal water is used to dry onions and garlic before it is injected back into the reservoir. Hawaii's geothermal plants provide about 20 percent of the electricity used on the Big Island. And the world's largest single geothermal power plant, 185 MW, is nearing construction near southern California's Salton Sea.

The Philippines and Indonesia have abundant geothermal resources. Geothermal generates about one-fourth of the electricity in the Philippines, making this country the second largest user of geothermal electricity in the world (after the United States). Italy was the site of the first geothermal power development. Its beautiful dry steam field of Larderello, developed in 1904, is still generating electricity today. Other places with large geothermal power developments include Mexico, Iceland, New Zealand, Japan, and several Central American countries.

"THE GEYSERS"

A geothermal field in northern California is named "The Geysers" (though it has no geysers — only fumaroles). It once was the site of a famous resort — attracting hardy travelers the likes of Jack London and Teddy Roosevelt. Today it is the world's largest single source of geothermal electrical power. Even its reservoir is rare, being one of the few in the world that produce steam (rather than mostly water). After over 50 years, The Geysers' 21 power plants still reliably generate enough electricity to power a city the size of San Francisco — about 900 MW.

To top it off, cities in Lake and Sonoma Counties are piping their cleaned wastewater many miles to The Geysers and injecting it deep into the geothermal reservoir. This practice helps sustain the productive life of the reservoir for electicity production while providing nearby cities with environmentally safe wastewater disposal.

Geothermal power plants at The Geysers in California

HOT LOCATIONS

The edges of the continents that surround the Pacific Ocean (the Pacific "Ring of Fire") are prone to earthquakes and volcanoes and have some of the best geothermal resources in the world. This includes the western part of North, Central, and South America; New Zealand; Indonesia; the Philippines; Japan; and Kamchatka (eastern Russia). These countries all have coastlines that sit on or near the boundaries of tectonic plates. Some of the other prime geothermal locations include Iceland, Italy, the Rift Valley of Africa, and Hawaii.

This map shows the edges of the tectonic plates (they all have names) that form the "Ring of Fire." Most countries in this area have lots of geothermal energy.

Enhanced Geothermal Systems:
New Energy from "Engineered" Reservoirs

There are many places underground where the rock is really hot but doesn't naturally contain much water. Researchers are working on ways to pump water down into this hot rock, creating "engineered" geothermal reservoirs. Called "enhanced geothermal systems" (EGS) or (less often) "hot dry rock," this method involves drilling a well into the hot rock and injecting high-pressure cold water to expand natural cracks (fractures) or to make new ones. Then more water is pumped down into the fractured rock. The heat from the rock transfers to the water, and the now-hot water is pumped up a separate well to generate electricity in a binary power plant.

A U.S. project at Los Alamos, New Mexico, first demonstrated that EGS can work. Japan, France, Germany, Switzerland, Australia, and other countries are also working on this method. In the United States and elsewhere, similar processes are being adapted to boost the production of already-developed natural geothermal reservoirs.

HOT ENERGY FOR A COLD COUNTRY

Iceland is such an active geothermal area that hot springs occasionally bubble up right into people's living rooms! People in this cool-weather country make really good use of their abundant geothermal energy resource. They use it for everything from heating homes, offices, and greenhouses to warming swimming pools and generating electricity. In the middle of winter, it is not uncommon to see people soaking in the steamy hot pool found right outside a geothermal power plant.

Some spas in Iceland are located right next to geothermal power plants.

CONSIDERATIONS

■ Geothermal power plants produce no smoke. What comes out the top of a geothermal plant cooling tower is steam (water vapor) with only trace amounts of natural minerals and gases from the geothermal reservoir. Flash and dry steam plants produce only a small fraction of air emissions compared to fossil fuel plants. (See page 135.) Binary power plants have virtually no emissions.

■ Geothermal power plants use very little land compared to conventional energy resources and can share the land with wildlife or grazing herds of cattle. They operate successfully and safely in sensitive habitats, in the midst of crops, and in forested recreation areas. However, they must be built at the site of the geothermal reservoir, so there is not much flexibility in choosing a plant location. Some locales may also have competing recreational or other uses.

■ Geothermal wells are sealed with steel casing, cemented to the sides of the well along their length. The casing protects shallow, cold groundwater aquifers from mixing with the deeper geothermal reservoir waters. This way the cold groundwater doesn't get into the hot geothermal reservoir and the geothermal water doesn't mix with potential sources of drinking water.

(continued)

CONSIDERATIONS (continued)

- Geothermal water contains varying concentrations of dissolved minerals and salts. Sometimes the minerals are extracted and put to good use. Examples are zinc (for electronics and for making alloys such as bronze and brass) and silica. At reservoirs with higher concentrations, advanced geothermal technology keeps the salty, mineralized water from clogging and corroding power plant equipment.

- Most geothermal reservoirs contain varying amounts of dissolved gases such as hydrogen sulfide. This gas smells bad (like rotten eggs), even at very low concentrations, and is toxic at high concentrations. Modern geothermal technology ensures that geothermal power plants capture these gases before they go into the air. Some gas removal processes can produce sulfur for use in fertilizers.

- Geothermal reservoirs must be carefully managed so that the steam and hot water are produced no faster than they can be naturally replenished or supplemented.

- Geothermal resources are generally developed in areas of high seismic activity, i.e., in earthquake country. Critics point out that some geothermal development — particularly with enhanced geothermal systems — can cause small earthquakes. Regulators are examining this issue where geothermal projects are near residential areas.

- Geothermal power plants run day and night, so they provide reliable baseload electricity. Most can increase their output of electricity to provide more power at times of greater demand. But geothermal power plants can't be used exclusively for peaking power; if geothermal wells were turned off and on repeatedly, expansion and contraction (caused by heating and cooling) would damage the wells.

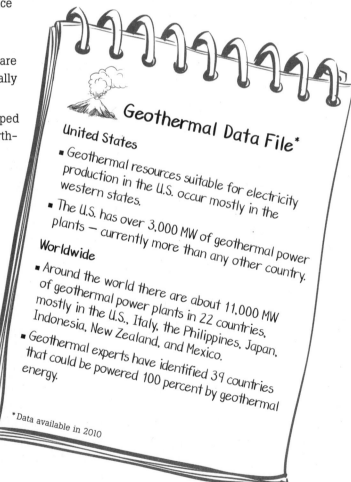

Geothermal Data File*

United States

- Geothermal resources suitable for electricity production in the U.S. occur mostly in the western states.
- The U.S. has over 3,000 MW of geothermal power plants — currently more than any other country.

Worldwide

- Around the world there are about 11,000 MW of geothermal power plants in 22 countries, mostly in the U.S., Italy, the Philippines, Japan, Indonesia, New Zealand, and Mexico.
- Geothermal experts have identified 39 countries that could be powered 100 percent by geothermal energy.

*Data available in 2010

Creating Electricity from Geothermal Resources

Finding Geothermal Resources

Abundant geothermal resources occur in the countries bordering the Pacific — the "Ring of Fire." Geologists explore volcanic regions like this steaming hillside in El Hoyo, Nicaragua, and other regions of the world where resources can occur even with no volcanoes or other surface evidence. *(Photo courtesy of Trans-Pacific Geothermal Corporation)*

Getting to the Geothermal Reservoir

If geologists find encouraging signs of deep heat, wells will be drilled. Drill rigs can be small water-well truck-mounted rigs, or large drill rigs like these, erected onsite. Geothermal wells can be drilled over two miles deep. *(Photo courtesy of Geothermal Education Office, Tiburon, CA)*

Testing a Geothermal Well

If a geothermal reservoir is discovered, the well is flow-tested to determine the pressure, temperature, and chemistry of the reservoir. Large "separators" can be used to safely control the hot steam, such as for this well test at the Blue Mountain power project in Nevada. *(Photo courtesy of Nevada Geothermal Power, Vancouver, Canada)*

Power Around the Clock

This turbine generator works fine outdoors at a geothermal project in California's Imperial Valley. Geothermal power plants provide baseload power, so these turbine generators operate 24 hours a day. *(Photo courtesy of Geothermal Education Office, Tiburon, CA)*

Creating Electricity from Geothermal Resources

World-Record Power
The Geysers steam field in northern California is the largest producing geothermal development in the world, with 21 power plants. Those white plumes you see are steam (water vapor); geothermal plants do not burn fuel or produce smoke. *(Photo courtesy of Calpine Corporation, San Jose, CA)*

Geothermal Plumbing
This flash steam plant is in East Mesa, California. Geothermal flash technology was invented in New Zealand. You can see the huge pipes bringing hot water and steam to the power plant, where it spins turbine generators. The used water is returned to the reservoir. *(Photo courtesy of Geothermal Education Office, Tiburon, CA)*

Compact Model
This small binary power plant is in Fang, Thailand. By using a heat exchanger, binary technology creates electricity from lower temperature reservoirs. Worldwide, developers are working on ways to produce power at ever lower temperatures.
(Photo courtesy of Geo-Heat Center, Oregon Institute of Technology)

Generation without Emissions
This air-cooled binary geothermal power plant nestles in the mountains at Mammoth Lakes, California. It is emission-free and consumes no water or chemicals. *(Photo courtesy of Ormat Technologies, Inc., Reno, NV)*

Renewable Energy Source:
HYDROPOWER

TERMS IN GLOSSARY

flow
head
headrace
horsepower
impoundment
micro-hydro
penstock
pumped storage
run-of-river (diversion)
tailrace
water cycle

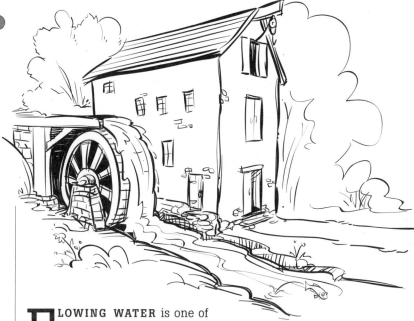

FLOWING WATER is one of nature's most powerful forces. Humans began harnessing this energy force several thousand years ago. By the first century B.C., waterwheels were working in many parts of the world, including Greece. (In fact, the term hydro comes from an ancient Greek word for water.) For centuries waterwheels in many countries provided the energy to grind grain and saw lumber. By the 1700s, more than 10,000 waterwheels were hard at work in colonial New England alone.

During the Industrial Revolution, waterwheels were also used to run textile mills and other factories. By the late 1800s water turbines were driving a new device – the generator – to produce electricity. Before the end of that century several commercial water-driven electrical stations were operating, including the largest at Niagara Falls, New York. The era of hydroelectric power was born.

THE HYDROPOWER RESOURCE

The hydropower resource is the force of flowing water, provided to us naturally by the earth's water cycle and by gravity. The force of the flow of a medium-size river is equal to several million horsepower. (One million horsepower, if converted to electricity, would equal the power of 746 MW.) You can imagine the appeal when this much force can be put to work driving waterwheels or water turbines.

REMINDER

W = watt
kW = kilowatt = 1,000 watts
MW = megawatt = 1,000 kilowatts
1 megawatt can serve about 1,000 homes in the United States.

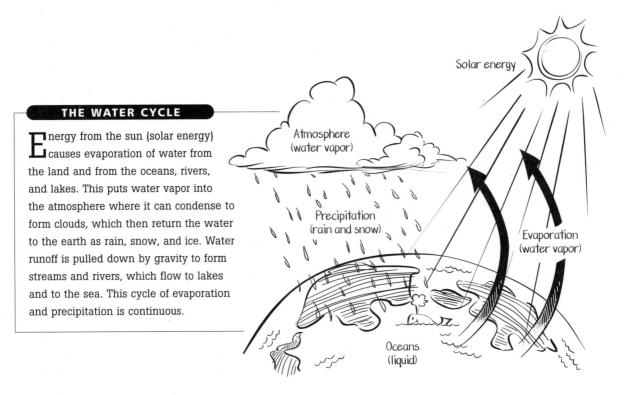

THE WATER CYCLE

Energy from the sun (solar energy) causes evaporation of water from the land and from the oceans, rivers, and lakes. This puts water vapor into the atmosphere where it can condense to form clouds, which then return the water to the earth as rain, snow, and ice. Water runoff is pulled down by gravity to form streams and rivers, which flow to lakes and to the sea. This cycle of evaporation and precipitation is continuous.

The Steeper the Better

The amount of force that water can impart depends on two factors: the head — the vertical distance the water falls; and the flow — the volume (amount or mass) of the water per second. The greater the head and the flow, the more water energy is available. So hydropower systems work best with a steep drop (high head) and a large flow. One gallon (3.8 liters) of water falling 100 feet (30 meters) in one second can generate about 1 kW of electric power. No wonder waterfalls, with their naturally steep drops, were chosen as the sites for the world's first hydroelectric power plants.

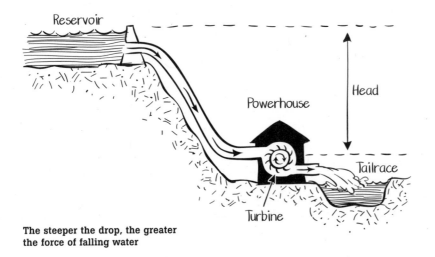

**The steeper the drop, the greater
the force of falling water**

GENERATING ELECTRICITY FROM HYDROPOWER RESOURCES

All hydropower plants, large or small, use a water turbine and a generator to produce electricity. The water turbine is at the heart of any hydroelectric system. Resembling its wooden ancestor, the waterwheel, it is far more streamlined and spins much faster. Interestingly, the first model is still in wide use. Its bucket-like metal paddles are enclosed in a shell into which the water flows. Today's water turbines are designed for maximum efficiency. They come in many shapes and sizes to work with varying conditions of head and flow. Hydropower generators resemble those used in many other electric power plants.

Most hydropower systems use some type of water passageway, channel, or pipe, called a penstock. The passageway concentrates the water's force by increasing the pressure as it approaches the turbine. That force turns one or more turbine-generators, which are usually enclosed in a powerhouse (to protect the equipment and to make maintenance possible). Water leaving the turbines is channeled downstream through a tailrace, back to the river.

HARD-WORKING WATER

The ambitious Big Creek hydropower project in California, begun in the early 1900s, now sends the water of Big Creek through a series of dams, lakes, tunnels, and powerhouses — all built into the steep mountainsides of the Sierra Nevada between Yosemite and Sequoia national parks. Nine powerhouses have been added, which together generate over 1,000 MW of electricity, prompting some to call this river system the "hardest working water in the world."

POWER SKETCH: Power in Paradise

Members of a family living in the hilly rainforest many miles from Quito, Ecuador, have always treasured their lush, natural environment. After many years of roughing it, they wanted to enjoy a few conveniences that required electricity. But they lived far from power plants and transmission lines. They solved this dilemma by installing a small hydroelectric system that uses water diverted from a fast-flowing stream on their property. This "run-of-river" system does not disrupt the flow of the river feeding the waterfall and pool below. It generates enough electricity to run a small refrigerator and electric lights. It even provides power to run a computer, which is used for their exotic plant-seed business. The forest has almost covered the power-generating equipment with foliage, so they enjoy the convenience of electricity without disturbing the beauty of their little piece of paradise.

Two Common Hydropower Systems

There are basically two ways that hydropower facilities use the force of flowing water: storage systems and run-of-river (diversion) systems.

Storage Hydropower Systems. The hydropower plants we're most used to seeing are called "storage" hydropower plants. These plants use a dam to hold back water, creating a reservoir and an artificially steep drop (high head). The dam is placed across a river, causing it to back up to form a reservoir or lake. The water is held back until it is needed. When released, it flows down through the penstocks to turbines in the powerhouse below. After the water passes through the turbines it is discharged into the river.

U.S. hydropower power plants produce almost 95,000 MW of power, much of which currently comes from storage hydropower facilities. About 60 percent of all electricity used in the Pacific Northwest is generated from hydropower. The largest storage hydroelectric facilities in America are found in this region. The Grand Coulee Dam, on the Columbia River, alone produces over 6,000 MW, making it the fifth largest storage hydropower facility in the world.

STORE NOW, USE LATER

Some dams use a pumped storage system to move water between an upper and a lower reservoir. During times of peak electricity demand, water is released from the upper reservoir to generate electricity and ends up in the lower reservoir. When electricity is plentiful, and this plant is not needed, electricity generated elsewhere is used. An example of this type of system is the Eastwood power plant in the Sierras of California.

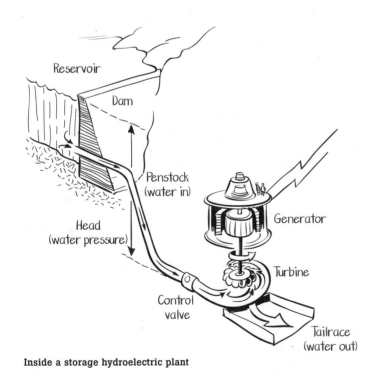

Inside a storage hydroelectric plant

Hydropower projects have had multiple roles in American history. With hydropower, the Tennessee Valley Authority (TVA) brought electricity to many rural homes and farms — the first time in the 1930s. Hoover Dam, near the Grand Canyon, helps control flooding along the mighty Colorado River and provides power to the southwest. The U.S. Bureau of Reclamation oversees multipurpose hydro projects nationwide for flood control, recreation, irrigation, and electricity.

China can claim the world's largest hydropower project. This project, on the Yangtze River, is designed to control flooding while producing an anticipated 22,500 MW of electricity. The second largest hydropower plant in the world is the Itaipú hydropower plant, which sits on the border of Brazil and Paraguay and produces over 12,000 MW of power. In fact, Brazil is the third largest producer of hydro-electricity in the world. Only Canada and the United States generate more. Western Europe, Japan, and Russia are also top hydropower producers.

CHUTES AND LADDERS

There are a number of ways to avoid damage to fish caused by large storage hydropower plants. Innovative methods include fish ladders for adult salmon migrating upstream to spawn, flashing lights to alert night-migrating fish, screens to shield the turbines, and surface collectors that guide juvenile fish through chutes that go around the project.

Run-of-River (Diversion). Run-of-river systems are today's hydro-power systems of choice, because they are designed to maintain the natural flow of a river and therefore are more wildlife-friendly than storage projects.

With these systems, the river generally continues to run its natural course while some of its water is directed off, or diverted, into a penstock. Once the diverted water has done its work, it is sent back to join the river through a tailrace. There are a number of ways this is accomplished.

In the simplest type of facility, a small dam directs water into a powerhouse at one end of the dam. After spinning the turbines, the water returns directly to the river. Sometimes the powerhouse is located further down the river at the end of a short canal, or headrace, to develop more head. In another type of facility, a penstock takes a steeper, more direct path than the river, rejoining the river down-stream. There are also systems that combine these elements.

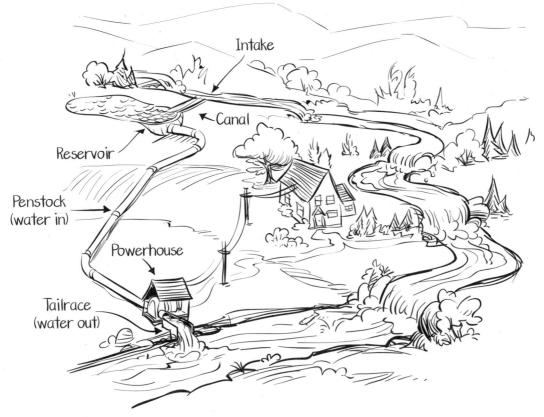

A diverted run-of-river system

One method that is favored in scenic areas uses a tunnel that is cut through rock alongside a steep drop. Nothing on the surface is disrupted as most of the system is placed underground. Snoqualmie Falls, providing the first hydropower to Seattle, Washington, was originally built that way. The Tazimina Hydroelectric Project near Anchorage, Alaska, uses the long vertical drop of a vigorous waterfall without marring its flow or its rugged surroundings. Some of the upper river's flow is diverted through a vertical pipe installed in the rocky cliff alongside the waterfall. The water rushes down the pipe to turbines in a powerhouse below and then rejoins the river's main flow.

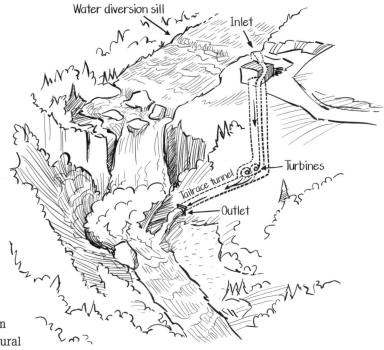

This run-of-river hydroelectric project in Tazimina, Alaska, does little to disrupt its natural surroundings.

Run-of-river hydropower is useful in many places. Because there is no large dam or reservoir, it limits disturbance to the natural setting. Also, it can provide electrical power for people living far from transmission lines. In the Gold Rush country of the western United States, some small run-of-river plants were built where water had originally been diverted by miners to "wash" gold out of gravel.

Sometimes storage and run-of-river systems are combined. For example, at Bishop Creek Hydropower Project in California's eastern Sierra Nevada, spring runoff from melting snow is collected in two reservoirs, built to prevent flooding of the Bishop Creek area below. Throughout the year, a moderate amount of water is released into the creek, where it eventually spills through four run-of-river powerhouses. The first use of this hydropower was to provide electricity for gold and silver mining in the early 1900s. These powerhouses still provide plenty of electricity today.

There are hundreds of run-of-river projects in the United States, and most new hydro projects being built in the U.S. are run-of-river. Many are hailed for preserving a river's flow while providing electrical power. One project is located just off the Mississippi River near Vidalia, Louisiana. It maintains the flow of the "Mighty Mississippi" (a main artery for transportation), helps control flooding, and supplies electricity.

Many small run-of-river projects are now being awarded coveted "low impact" certification for producing hydropower without disturbing the local environment. One example is the Lower Robertson hydro project on the Ashuelot River in New Hampshire. Many other facilities have also received this award, including one at Falls Creek in the Willamette National Forest of Oregon.

Worldwide, there is great interest in run-of-river projects for both remote and grid-connected areas. For example, mountainous Nepal, which features over 6,000 rivers and streams, is interested in using these systems to provide rural villages with electricity. China has plans to provide up to 75 million people with electricity from diversion projects. Many hilly areas in Europe are already dotted with run-of-river hydro systems.

Project Size

Today's hydropower systems range from those that provide energy for one home to mammoth, multi-megawatt installations. A common definition of a large-scale hydropower facility ("large hydro") is one that generates more than 30 MW of electricity. Most large-scale hydropower installations use a storage system, which creates the greatest environmental concerns. (See "Considerations," next page.)

Small-scale hydropower projects are often divided into small-scale hydro and micro-hydro. Small-scale projects range from 100 kW to 30 MW, while micro-hydro projects usually produce 100 kW or less. Small-scale projects can be run-of-river or storage-type facilities. Micro-hydropower systems are run-of-river, considered to be gentler on the environment.

Most experts say that nearly all new hydropower facilities will be small-scale and/or run-of-river facilities. Many potential locations exist for these types of power plants. Also, some older small-scale hydro projects that were shut down in the 1930s and '40s have recently been reopened. For example, a small hydro facility operated by Cornell University at Fall Creek in Ithaca, New York, was reopened in the 1980s after a 30-year shut-down. It is once again supplying 1.5 MW of power to the university, a small but eloquent testimonial to the value of using a local, renewable energy resource.

CONSIDERATIONS

■ Hydropower produces no air pollution. It is very efficient and — once installed — inexpensive. Storage hydropower plants run reliably all year long, providing good baseload power, though long-term drought can reduce overall capacity. They can be started up relatively fast when needed for peaking power.

■ Run-of-river hydropower systems are considered by many to be the preferred hydropower technology because they are easier on the environment. Run-of-river projects could contribute a significant amount of electricity worldwide.

■ A drawback of run-of-river systems is that the flow in the rivers and streams fluctuates by season, and in low rainfall or drought years, less electricity can be produced. Low rainfall levels can also reduce seasonal electricity output of storage systems.

■ Large-scale storage hydropower projects are expensive to build, but can provide many megawatts of electricity to an area for decades.

■ The installation of a dam across a river for a large-scale storage project can cause the river water to back up over hundreds to thousands of acres, flooding significant land areas. There are impacts to water quality, fish, and wildlife. The flooding can also ruin important cultural, religious, and archeological sites and can displace hundreds or even thousands of people. The large Three Gorges project in China has created a lake 400 miles (644 kilometers) long and required the relocation of 1.2 million people.

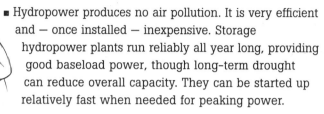

Hydropower Data File*

United States

■ The United States gets about 6 percent of its electricity from hydropower.

■ Washington, Oregon, California, and New York are the states currently producing the largest amount of hydroelectricity.

Worldwide

■ About 20 percent of the world's electricity is generated by hydropower.

■ Sixteen countries get about 90 percent of their electricity from hydropower, including Norway, Brazil, Paraguay, Albania, Ethiopia, Ghana, and Zambia.

* Data available in 2010

(continued)

CONSIDERATIONS (continued)

- Most dams serve multiple purposes, like flood control, water supply, transportation, and recreation. In fact, fewer than 3 percent of the dams in the United States have hydropower plants, and for many that do, electricity is a secondary purpose.

- Currently, a number of large projects around the world have been canceled or placed on hold due to public concern about the environmental impacts of the dams. In recent years, some consideration has been given to removing dams in highly sensitive areas. Some smaller, older dams in sensitive areas of the U.S. Pacific Northwest have already been removed.

Creating Electricity from Hydropower Resources

Run-of-River at Ashuelot, New Hampshire
This is a low impact certified project on the Ashuelot River. Its three turbine generators, with a capacity of 960 kilowatts, are deep under the water. *(Photo courtesy of Bob King, Ashuelot River Hydro, Inc., Keene, NH)*

Oregon Fish Ladder
The Vine Street Hydro project in Oregon features a fish ladder to bypass the 6-foot high concrete diversion dam. A fish screen deters migratory fish from straying into the canal. *(Photo courtesy of the City of Albany, OR)*

Penstock
A 10-ft. diameter penstock at the Millpond Project in New York diverts water from Catskill Creek. The penstock run carries water directly to the power-house, through the turbine, and right back into the river. *(Photo courtesy of Bob King, Ashuelot River Hydro, Inc., Keene, NH)*

Run-of-River in Maine
This is a 15 MW hydro project in Kennebec, Maine. The turbine generators are inside the little power-house on the right. The force of the water going from the higher to the lower elevation spins the turbine. *(Photo courtesy of Brookfield Renewable Power, Marlborough, MA)*

Creating Electricity from Hydropower Resources

Low Impact in New York
This run-of-river project is on the Racquette River. In this photo you can see the "head pond," a small dam, a little power house with four turbines, and the tail race (at a lower elevation), where the water, having done its job, rejoins the river. *(Photo courtesy of Brookfield Renewable Power, Marlborough, MA)*

Sharing Nature in Yosemite National Park
A curved gravity dam captures the snowmelt on the Tuolumne River, creating the Hetch Hetchy Reservoir. This storage reservoir provides water and electricity to San Francisco hundreds of miles away. Generating facilities and transmission lines — concealed to protect the valley's famous scenery — produce about 500 MW of electricity. *(Photo from Wikipedia)*

Pumped Storage
These are pumped storage hydropower facilities. Above left is the Jack Cockwell pumped storage hydro facility on the beautiful Deerfield River in Massachusetts. The facility on the right is in Muddy Run, Pennsylvania. During off peak hours the power house pumps water to the upper reservoir; during times of peak power demand the water is directed down through the turbines to generate electricity. *(Cockwell photo courtesy of Brookfield Renewable Power, Marlborough, MA; Muddy Run photo courtesy of Exelon Corporation, Chicago, IL)*

Renewable Energy Source: OCEAN

TERMS IN GLOSSARY

aquafarming

barrage

current

ebb

estuary

high and low tide

marine current

Ocean Thermal Energy Conversion (OTEC)

one-way marine current

sluice

strait

tidal current

tidal fence

tidal power plant

SINCE EARLIEST TIMES, the ocean has been a vast resource for travel, food, pearls, minerals, oil, and much more. Some say that the ocean is the last remaining frontier on earth. Much of the deep seafloor, with its many marvels, remains to be explored. And there's the lure of undiscovered shipwrecks and the riches they might contain. However, there is yet another ocean frontier that some think is much more valuable than buried treasure. This is the ocean's energy frontier, one that we are just beginning to understand.

THE OCEAN RESOURCE

Oceans create tremendous energy in the movement of their currents, tides, and waves. The oceans also store a vast amount of heat from the sun. With today's technology these ocean energy resources can be put to work generating electricity.

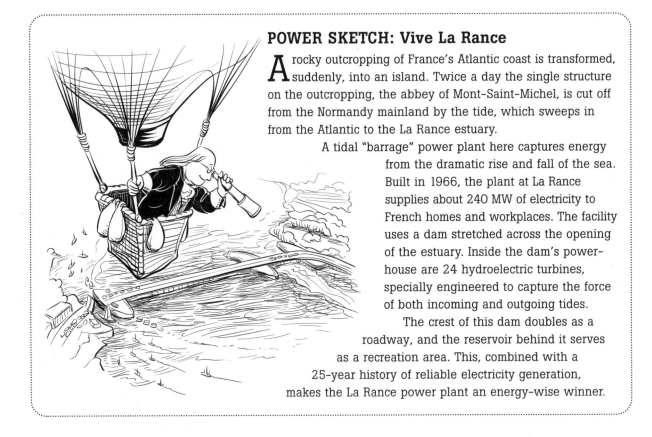

POWER SKETCH: Vive La Rance

A rocky outcropping of France's Atlantic coast is transformed, suddenly, into an island. Twice a day the single structure on the outcropping, the abbey of Mont-Saint-Michel, is cut off from the Normandy mainland by the tide, which sweeps in from the Atlantic to the La Rance estuary.

A tidal "barrage" power plant here captures energy from the dramatic rise and fall of the sea. Built in 1966, the plant at La Rance supplies about 240 MW of electricity to French homes and workplaces. The facility uses a dam stretched across the opening of the estuary. Inside the dam's powerhouse are 24 hydroelectric turbines, specially engineered to capture the force of both incoming and outgoing tides.

The crest of this dam doubles as a roadway, and the reservoir behind it serves as a recreation area. This, combined with a 25-year history of reliable electricity generation, makes the La Rance power plant an energy-wise winner.

Marine Tidal and Ocean Currents

There are two kinds of marine currents: two-way (tidal) currents, and one-way currents, also called ocean currents.

Two-way currents are the ocean tides, caused by gravitational pull of the moon and the sun. Each heavenly body pulls on the part of the ocean nearest to it, causing bulges in water height. As the earth rotates, those bulges created by the moon and the sun move in relation to the world's coastlines, pulling water into and away from the shore, concentrating kinetic energy in narrow coastal waterways. So the turning of the earth causes a moving pattern in the ocean: the level rises and falls, resulting in two high tides and two low tides every day in most parts of the world.

One-way currents are like massive "rivers" of ocean water flowing within the ocean for hundreds — sometimes thousands — of miles. Found close to the surface or deep within the ocean, one-way currents occur because of differences in ocean temperature, water density, and

REMINDER

W = watt
kW = kilowatt = 1,000 watts
MW = megawatt = 1,000 kilowatts
1 megawatt can serve about 1,000 homes in the United States.

topography from area to area. Some one-way currents are also driven by winds and the earth's rotation. One well known one-way current is the Gulf Stream, which carries warm water from the Gulf of Mexico to the northerly shores of Europe. The Gulf Stream is currently the focus of research on one-way ocean current technology.

Both two-way and one-way marine currents can be used to generate electricity. The key is a site with a forceful flow of ocean waters. Strong flows tend to occur within straits, between islands, and at entrances to large bays and harbors. In North America, the most promising tidal current locations are in the northern part of both coasts. Canada has the best opportunity to tap this renewable resource. Worldwide, high-energy marine current tidal sites are in the United Kingdom, Italy, Australia, India, Russia, Argentina, the Philippines, and Japan.

Waves

Wind force creates ocean waves. The circular motion of water molecules swirl with loads of kinetic energy. Waves, along with marine currents, are 800 times more powerful than wind! Each wave can travel a long distance without losing much energy. Wave energy is remarkably powerful, something that can be confirmed by anyone who has ridden a big wave (or been knocked down by one). Strong, steady winds will produce waves that travel up to 35 miles (56 kilometers) per hour. Storms, of course, produce the most dramatic waves, some of which tower over 100 feet (30 meters).

Wave energy systems can use the moving force of these waves in the open sea or along coastlines. Experts estimate that wave energy has the potential to provide enough electricity to match conventional hydropower production in the U.S. They also think the best spots could produce up to 65 MW per mile of coastline. These high-energy wave locations are generally on western coastlines facing the open sea, including California and the Pacific Northwest in North America, Chile, northern Europe, South Africa, Australia, and New Zealand. Potential for a 100 MW wave energy project on California's central coast is being studied.

The Ocean's Stored Heat

Most of the solar radiation that the earth receives falls on the world's oceans, where it is absorbed and stored as heat energy. Our oceans cover more than two-thirds of the globe, so you can imagine how much heat is stored in these vast waters. This heat energy is found in the upper surface waters. The sun's rays are not able to penetrate extremely deep water, but down to about 200 feet (61 meters) they can be an effective warming source. Deeper water, by contrast, is very cold — near freezing at depths greater than 3,000 feet (914 meters).

The temperature difference between sun-warmed surface water and near-freezing deep water is great enough to be used to produce electricity. The difference needs to be at least 36°F (22°C), so some of the best global locations are the sun-drenched ocean surfaces of the tropics. That includes Hawaii, Florida, and many other places, including island nations such as those of the southern Pacific and Indian oceans.

GENERATING ELECTRICITY FROM OCEAN RESOURCES

Marine Current Energy Systems

Some marine current systems aren't new. As long ago as the Middle Ages (1200-1500 A.D.), farmers built ponds to trap advancing seawater from rising tides. They would then use the moving water to power their mills when the tide flowed back out. Today, some older tidal power plants use a dam, or barrage, stretched across the opening of a bay or strait. At high tide, water rushes through openings, or sluices, into a reservoir, bay, or estuary. The water is held there until the tide drops (ebbs), and then it's let out through turbines that drive electrical generators. A variation of this technique is a system that uses generators working at both low (outgoing) and high (incoming) tides. This second type is what makes the La Rance facility work. (See "Power Sketch," page 74, and illustration, page 77.)

Most types of marine current technologies are still in the research stage, though several commercial facilities are now under development. One type is a tidal fence, a series of underwater turbines resembling a row of giant turnstiles, set up in any kind of channel — between two islands, for example. Far more common is a tidal turbine that works like an underwater wind turbine. Mostly submerged in the sea, these systems can also be grouped into undersea "energy farms."

The temperature differences in the ocean result in dramatically different habitats. Some ocean fish thrive in upper surface waters warmed by the sun. Others have adapted to life in deeper, much colder water.

A number of countries have been investing in marine tidal current energy. China began using tidal power in the mid-1950s. At one point as many as 40 small tidal power plants were generating electricity there. Most of these became outdated and were closed, but China still has at least 7 tidal power stations that altogether generate about 11 MW of power. A power plant located on the Bay of Fundy in Nova Scotia has been generating around 20 MW of tidal power since 1984. In the United Kingdom, several research groups are exploring various tidal current energy technologies. A Scottish company has installed two 300 kW tidal turbines in northern Scotland. Six 34 kW turbines are operating on the East River, right in New York City. Other countries looking at the use of marine current energy include India, Korea, Russia, Australia, the Philippines, the U.K., Canada, and the United States.

Wave Energy Devices

Shorelines that experience steady, powerful wave action are good places for stationary Wave Energy Devices. One type uses forced air to spin a turbine. Waves enter a column and force air past the turbine, making it spin. When the waves recede, the air pressure drops, also causing the turbine to spin. So electricity is generated as the waves enter the column and again as they recede. (See illustration next page.) Scotland has the world's first forced-air type of wave power plant, capable of producing 500 kW of electricity.

A different type of shoreline wave device uses a reservoir built into the bottom of a cliff. Waves are channeled into the reservoir, from which the water is released when needed to produce electricity. In a different variation, the water is moved up into an elevated reservoir using an anchored mechanical pump that runs on wave action.

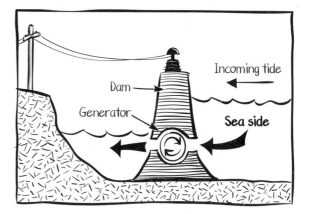

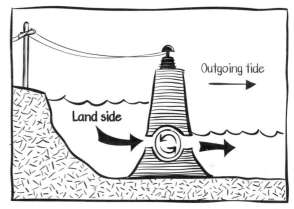

A tidal generation power plant that works at both high and low tides

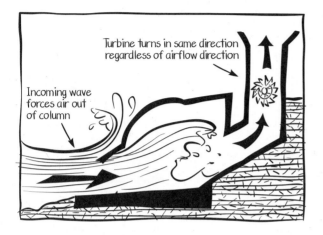

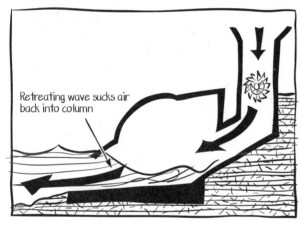

A forced air Wave Energy Device, called an oscillating water column

Far more common types of wave energy generators are offshore floating devices. These dot the ocean's surface, attached to the seafloor with anchors. Some types use the bobbing motion of the waves to cause pumps or pistons to move up and down, converting the force to spin the turbines. In other systems, the turbines are powered by the surge of the waves as they pass through. Some floating wave systems can also be grouped together, creating a "wave energy farm."

A wide variety of floating wave energy designs have been researched, with colorful names such as the "Clam," "Wave Dragon," and the "Nodding Duck." A design that has attracted a lot of media attention is named "Pelamis," after the Greek sea snake. Many wave energy systems have been developed in northern Europe, which experiences strong wave (and tidal) activity and is considered the world's leader in the ocean energy field.

Several wave energy projects are being explored in the United States, including one for the Kaneohe Marine Corps Base in Hawaii, one in Humboldt Bay in northern California, and the largest in Reedsport, Oregon. At Port Kembla, near Sydney, Australia, a forced air wave energy system will soon deliver electricity to almost 600 homes. Other countries investigating wave energy include China, India, Japan, New Zealand, Portugal, Indonesia, and Canada.

Nodding Duck

Ocean Thermal Energy

The temperature difference between warm surface seawater and cold deep seawater is put to work in Ocean Thermal Energy Conversion (OTEC) power plants. These can be located along the shoreline or on floating offshore platforms. There are two main types: closed cycle OTEC and open cycle OTEC.

Closed Cycle OTEC. In a closed cycle OTEC system, warm ocean water is piped into a heat exchanger that causes another liquid with a lower boiling point (the "working" liquid) to "flash" to vapor, or vaporize. (See "Heat Exchangers," page 53.) The force of the rapidly expanding vapor drives a turbine. Cold water from the ocean depths is brought up in a different set of pipes to cool the vapor, causing it to condense back into a liquid. This system's name comes from the fact that the working liquid is contained in a closed pipe system.

Open Cycle OTEC. In an open cycle OTEC system (see illustration on this page), the warm seawater itself is the working liquid. It is piped into a chamber in which the pressure has been reduced by creating a partial vacuum. The lower pressure allows the warm water to flash to steam. Cold seawater is used to condense the steam after it has passed through the turbine. An open cycle OTEC test plant in Hawaii successfully ran from 1993–1998. Its design reflected years of work done by several U.S. agencies and a university in Japan.

Closed cycle OTEC is most useful for producing large amounts of electricity and can be used with already existing turbine designs. Open cycle OTEC, which needs modified turbines, may be best suited for places such as island nations that need some fresh drinking water as well as electricity. In addition to the U.S. and Japan, other places studying OTEC include India, the Republic of Palau, and Fiji.

Ocean electricity-generating systems can turn salty seawater into fresh drinking and irrigation water. For example, an open cycle OTEC power plant in Hawaii at one time produced up to 7,000 gallons (24,498 liters) of fresh drinking water every day while generating electricity. And the McCabe Wave Pump, first designed 20 years ago to produce electricity, has been adapted to provide pumping power to produce 100,000 gallons (378,541 liters) of clean water daily from the seawater of the Shannon River Estuary in Ireland.

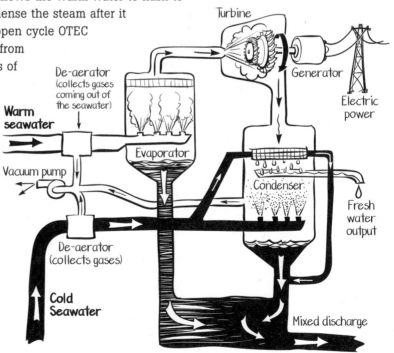

An open cycle Ocean Thermal Energy Conversion (OTEC) system

CONSIDERATIONS

- Until recently, the vast energy potential of our oceans had not been widely recognized. Luckily, we are now entering an era in which many different kinds of ocean energy technologies are being perfected and put to work.

- Wave and tidal energy systems produce no polluting emissions. OTEC produces a small amount of carbon dioxide, but only 4–7 percent of that produced by a traditional power plant.

- The tides, waves, one-way ocean currents, and thermal qualities of oceans are fairly predictable resources. They could produce significant amounts of baseload and peaking power. Tidal power plants are most useful near populated coastlines with high energy requirements.

- The effects of tidal power plant operations on the local environment depends on the plant location and the type of technology. Traditional barrage tidal systems have had some unavoidable environmental and visual impacts. The current trend is towards tidal energy systems that do not use a barrage. Still, many tidal turbines look like underwater wind turbines. Safety concerns center around fish, turtles, and whales, instead of birds and bats.

- At present, OTEC systems are much more expensive than power generating systems using most other energy sources, due primarily to the costs of installing and protecting pipes and equipment in the rigorous conditions of the ocean waters. OTEC is the least efficient of all ocean energy options, and it may only be viable at islands also seeking fresh water.

- Connecting offshore ocean energy projects to the on-shore electrical grid is a technical challenge.

Ocean Data File*

United States
- Though there are currently no commercial ocean energy power plants in the United States, several new demonstration projects are showing great promise.
- In the United States, parts of the Pacific Northwest and New England have significant tidal resources.

World
- The heat energy from the sun that is absorbed each day in the ocean has been compared to the energy stored in 250 billion barrels of oil.
- Approximately 300 megawatts of tidal energy is currently being produced worldwide.

*Data available in 2010

Creating Electricity from Ocean Resources

Scottish Oyster®
This wave energy converter has a simple hinged flap connected to the near-shore seabed; hence the "Oyster" name. Each passing wave moves the flap, driving hydraulic pistons to deliver high pressure water through a pipeline to an onshore electrical turbine. The device is intended for use in groups as wave farms. Most of the Oyster, when operating, is under water. *(Photos courtesy of Aquamarine Power, Edinburgh, Scotland)*

Pelamis®
Here is Pelamis, a wave energy converter at the European Marine Energy Centre in Orkney, Scotland. This model is used in the first commercial wave farm project, off northern Portugal. *(Photo courtesy of Pelamis Wave Energy Ltd., Edinburgh, Scotland)*

Caring for the Irish Environment
Designed to capture tidal energy, this open-centre turbine rests directly on the seabed. It is invisible from the surface. Its turbine blades turn from the force of the tides, and its slow-moving rotor minimizes risk to marine life. The regularity of the tides makes its output steady and predictable. *(Photo courtesy of Open Hydro Tidal Technologies, Dublin, Ireland)*

Creating Electricity from Ocean Resources

Riding the British Tidal Currents

SeaGen® is a turbine system that extracts energy from the ebbing and surging of tides. Comprised of two turbines mounted on either side of a structure, it resembles an under-water windmill. A convertor connects to a marine cable, which carries electricity to the grid onshore. In the photo on the right, note size comparison of SeaGen to the workers in the boat. *(Photos courtesy of Marine Current Turbines Limited, Bristol, UK)*

PowerBuoy® in Spain.
Ocean waves create utility-scale electrical power in the PowerBuoy. Upright on the sea floor, the floating piece moves up and down with the waves, capturing energy to drive the generator. Electricity is transmitted via an underwater power cable. The photo on the right shows PowerBuoy out of the water on its side. It's 12 feet in diameter and 55 feet long. *(Photos courtesy of Ocean Power Technologies, New Jersey)*

Capturing Currents in NY
Underwater free-flow systems can draw energy from the current in different settings — oceans, rivers, and man-made channels. Turbines like this one have been test-operated successfully in the East River right in New York City. *(Photo courtesy of Verdant Power, Inc., NY)*

Renewable Energy Source: SOLAR

TERMS IN GLOSSARY

array
central receiving tower
Concentrating Solar Power (CSP)
electromagnetic spectrum
heliostat
infrared
module
parabolic trough
photon
photoelectric effect
photovoltaic (PV)
radiant energy
semi-conductors
silicon
solar cell
solar dish engine
solar panel
spectrum
thin film PV
ultraviolet

WITH SOLAR ENERGY, THE SKY'S THE LIMIT. Our sun is the world's most widely used energy resource. Plants began capturing the sun's energy millions of years ago, and members of the animal kingdom have always basked in its warmth. Human dwellings have long included openings that let in the sun's light and heat. Glass windows were used as early as 79 A.D., as revealed in the archeological ruins of Pompeii and Herculaneum (Roman cities completely preserved under layers of ash from a volcanic eruption). Now, our use of windows to admit the sun's radiation is such a common practice that we don't even think about it. And today, with technology ranging from tiny solar cells to huge power plants shimmering with rows of curved mirrors, we use solar energy to make electricity.

THE SOLAR RESOURCE

We all know that our sun gives off radiating waves of heat and light energy. Without these, our planet would not have life. The sun's waves move rapidly — at the speed of light — as tiny bundles of energy called photons. These photons travel vast distances from the sun through the vacuum of space and bathe our planet with solar energy every day.

Shedding Light on the Solar Spectrum

All the sun's radiant energy waves form the solar electromagnetic spectrum. Forty-five percent of the radiation of the solar spectrum that reaches the earth's surface is visible light. The rest we do not see (although we can detect and measure it), yet it all delivers energy. For example, ultraviolet radiation, though we can't see it, can tan or burn our skin. And we're all familiar with the sun's infrared, or thermal (heat), radiation. Infrared radiation is what keeps the earth (and us) warm.

Some parts of the earth receive more solar radiation than others. In general, the areas at or near the equator receive the most. For example, the tropics get about two and a half times more infrared radiation than the north and south poles. However, any area that receives a steady supply of solar radiation, whether a little or a lot, can make use of the energy pouring in from our sun.

We use just a fraction of our enormous solar resource.* More energy from sunlight strikes Earth in one hour than all the energy consumed on the planet in a year.

GENERATING ELECTRICITY FROM SOLAR RESOURCES

In this section we discuss solar energy strictly as a source of electricity. In Chapter 5 we discuss direct (non-electric) uses of solar energy — active direct uses such as heating water, and passive direct uses such as designing sun-friendly homes.

Photovoltaics (PV)

In the 1950s, American engineers sought a method to power U.S. space satellites. They found it in a technology called photovoltaics (PV). We still use PV to energize orbiting satellites, space stations, and the Hubble telescope. Back on Earth, PV is widely used for everything from roadside call boxes to utility-scale power plants, though the most widespread application is for on-site power generation.

THIN FILM PV

Today there is a slimmer version of PV technology, something called thin film PV. Thin film PV can be used to replace some of the regular shingles on a building's rooftop. Operating in the same way that flat plate PV does, thin film shingles are as durable and protective as regular asphalt shingles. These solar shingles are textured to fit right in with the architectural design of buildings. Ultra-thin versions of thin film PV may also be applied to windows and the sides of skyscrapers.

---

*Statistics buffs, take note: The total amount of solar radiation received by the earth is 1.73×10^{17} watts at any one time. This is enough to warm our entire globe, fuel all of the earth's photosynthesizing plants, and create global climatic systems that drive the winds, the waves, and the water cycle.

POWER SKETCH: Lighting the Way on a Foggy Day

On foggy days off the coast of Ventura, California, a lone lighthouse shines its lights and sounds its foghorn for maritime travelers. Though far from the mainland's electrical connections, the Anacapa Island lighthouse operates entirely on electricity coming from a large group of solar panels. This electricity converted from sunlight also charges batteries to operate the lights even when the sun doesn't shine. Anacapa is part of the Channel Islands National Park system, a series of islands for which diesel generators once provided the electricity.* Now, instead, dozens of solar panels are powering operations around the islands, including a naval installation on Santa Cruz Island.

*Some of the generators remain, now using cleaner-burning biodiesel, but only as a back-up.

In PV, photons of sunlight react with specially designed semi-conductors in a process that results in electricity. *Photo* means light; *voltaic* refers to the electrical current. The smallest unit is a photovoltaic cell, made of wafer-thin layers that react to sunlight to create electricity. The most common photovoltaic material in use today is silicon, either in crystalline form or thin films, but other materials are being developed (see "Inside a Solar Cell," next page).

Usually, about 40 solar cells are wired together into a module, or panel. A number of PV modules, depending on the desired amount of electricity to be produced, are then wired together into a PV system known as a solar array to provide electricity for a household or business. A typical household array might have between 10–25 modules, depending on the size of the home. Hundreds of arrays (known as an array field) are grouped together for use by a large commercial or industrial facility or by a utility.

PV systems can be stand-alone (not connected to electric transmission lines) or grid-connected. With grid-connected PV systems some homeowners can sell their extra electricity to their local utility (see chapter 5).

PV panels on the roof of a house

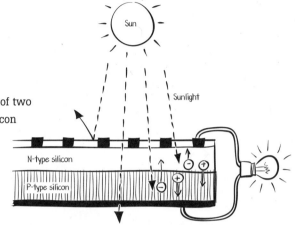

INSIDE A SOLAR CELL

A solar, or photovoltaic, cell is a "sandwich" made of two slightly different, super-thin layers of treated silicon crystals. When a photon of light from the sun strikes a solar cell, it frees electrons from some of the atoms of the treated silicon materials. These freed electrons zoom away from their "parent" atoms, leaving behind "holes." Because of the types of materials found in each layer, the electrons, which are negatively charged, tend to collect in what's called the N-layer (N for negative), and the positively charged "holes" collect in the P-layer (P for positive). When wires connect the two layers, electrons flow through the wire circuit in an orderly way. This is because negative and positive charges attract each other. This flow creates a current of electricity. (The freeing of electrons in solar cells by photons of light from the sun is called the "photoelectric effect." Albert Einstein won a Nobel Prize for describing it.)

Stand-alone PV. Photovoltaic systems are very handy for remote locations where transmission and distribution lines are not environmentally desirable or financially practical. These stand-alone systems are useful for lighting highway signs (energy is stored in batteries for use at night), roadside call boxes, and unmanned research installations in remote areas. They are also frequently found in rural areas or in national parks for lighting, battery charging, driving electric motors, water pumping, and more. The airport at Glen Canyon National Recreation Area, Utah, for example, is powered entirely by PV. Pinnacles National Monument in California uses solar cells for all operations including the ranger station, residences, and campground.

**PV array on a building at
Pinnacles National Monument**

Globally, stand-alone PV is providing electricity in many developing areas that have no widespread transmission lines. Indonesia, a nation of 17,000 islands, is turning to PV electricity rather than trying to connect all the islands with transmission wires. India has supplied hundreds of complete PV "kits" (called Solar Home Systems) to its rural villages. These include everything needed to light up a small home, including solar panels, wiring, and even the lights themselves. In Morocco, on the edge of the North African desert, solar panels have often been found at bazaars, where they are sold right alongside exotic Moroccan rugs and tin ware.

Grid-connected PV. Grid-connected PV systems range from small rooftop home set-ups to large PV power plants. Today, thousands of public and private buildings are being fitted with PV. Many businesses, such as warehouse-type stores, are making use of their expansive rooftops to install solar panels and now represent the largest segment of the U.S. PV market. Hundreds of utilities are including PV in their operations. The Sacramento Municipal Utility District in California, for instance, has more than 1,100 PV systems (including 800 to 900 homes with PV roofs) that together can produce about 11 MW. The first neighborhood to put PV on the roofs of all of its homes is in Gardner, Massachusetts. These were installed in the 1980s.

Worldwide Use of PV

In the U.S., California is the largest user of grid-connected PV. New Jersey, Illinois, Ohio, New York, Arizona, Texas, and Colorado are also making wide use of grid-connected PV systems. Globally, millions of small PV systems are in use. PV power plants that generate at least 1 MW or more of solar electricity are operating in the United States, Germany, Spain, Italy, India, and Japan. In fact, in Japan over 40,000 PV systems were installed in the year 2002 alone. The fastest growth has been in Germany, where from 1990 to 2001 PV use increased 58 percent. By the end of 2008, total PV capacity globally reached more than 15,000 MW. The three leading countries are Germany, Japan and the U.S. Together, these countries produce almost 90 percent of the world's total PV power.

REMINDER

W = watt
kW = kilowatt = 1,000 watts
MW = megawatt = 1,000 kilowatts
Using solar energy, 1 megawatt serves about 500 homes in the U.S.

Solar Thermal: Concentrating Solar Power (CSP)

CSP systems use mirrors to concentrate the energy from the sun to heat liquids held in pipes and tanks. Using a heat exchanger (see "Heat Exchangers," page 53) these hot liquids are vaporized to drive a turbine to generate electricity. CSP works best with a clear, dry sky and a high concentration of the sun's rays. In the United States, the sunny southwestern states have been actively exploring this technology. Sun-drenched areas such as India, Morocco, Egypt, and Mexico are also very interested in CSP. CSP systems range from small individual 5 kW units suitable for a remote facility, to huge, utility-scale systems that can produce 200 MW or more.

All CSP systems have two main parts, one that concentrates solar energy's heat, and another that converts this heat energy to electricity. The three main types of CSP systems are solar dish engines, parabolic troughs, and central towers (central receivers).

Solar Dish Engines. Though not yet widely used, solar dish engines, presently under development, may turn out to be the best option for remote and rural locations. They are composed of two parts: a curved (parabolic) mirror that concentrates the sun's heat, and a Stirling engine (see "A Stirling Idea," page 21) that uses the heat to generate electricity. Dish engines can be used individually, providing between 5 to 25 kW, which is enough power for a farm or village, or can be combined for large-scale, grid-connected operations.

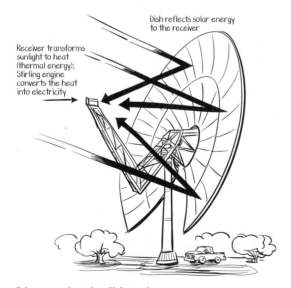

Dish reflects solar energy to the receiver

Receiver transforms sunlight to heat (thermal energy); Stirling engine converts the heat into electricity

A large-scale solar dish engine

STORING SOLAR ENERGY

Batteries aren't just for *supplying* electrical energy, but also for *storing* electrical energy. This storage capacity is very useful in solar energy systems, since sunlight isn't always available. The same process that provides an electric charge in a battery will also work in reverse. The inflow of electrons from the solar cell causes chemical substances in the battery to recombine and change. This "stores" the energy by charging the battery. When electricity is needed, the battery is activated, causing another chemical reaction that results in a flow of electrons — generating an electrical current.

There are other ways to store the sun's energy. One of these is collecting and holding the sun's heat. Concentrating Solar Power (CSP) systems use materials that hold a large amount of heat and then release it very slowly. One material often used is molten salt, which reaches very high temperatures and retains the heat for long periods of time. When the sun stops shining, this "set-aside heat" continues to run the CSP equipment that generates electricity. Salt is also used to hold the sun's heat in large outdoor pools called solar ponds.

Parabolic Troughs. Parabolic troughs are long, trough-shaped reflectors that focus the sun's energy on a pipe running along the mirror's curve. The concentrated heat warms up an oil flowing through the pipe. Heat energy from the oil is cycled through a heat exchanger (see "Heat Exchangers," page 53) to boil water to create the steam that drives the turbine. Most parabolic troughs are large, but smaller "roof-top" models are also being developed. Spain is researching a very promising system that boils water for steam directly at the parabolic trough.

Parabolic troughs rotate from side to side, so they can track the sun as it moves from east to west. They are normally located in many parallel rows. The Mojave Desert is home to the world's largest parabolic trough facility, where nine power plants feed around 350 MW of electricity to southern California homes and businesses.

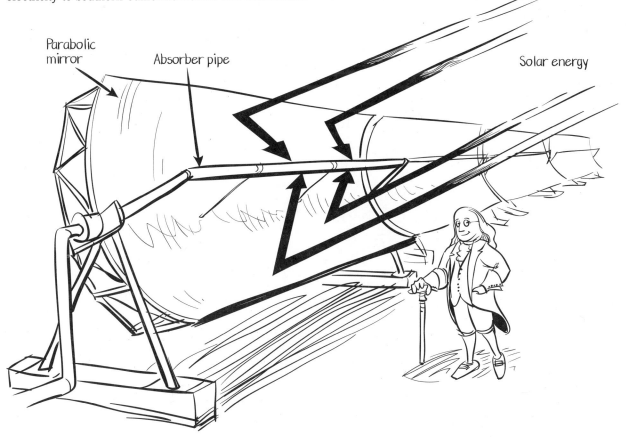

Parabolic mirror

Absorber pipe

Solar energy

A parabolic trough uses pipes containing clear oil that absorbs heat reflected off the trough. The heat from the oil flows through a heat exchanger to heat water to make steam for electrical generation.

Central Receiving Towers. Central receiving towers, often called "power towers," are tall structures with a boiler on top that houses a liquid suitable for heating, such as water (as shown below), molten salt, or liquid metal. Surrounding the tower are many rows of mirrors, called heliostats, which turn to face the sun and focus its rays onto the tower boiler throughout the day. The concentrated sunlight from these mirrors heats the liquid to as high as 2,700° F (1,482°C). This produces boiling water, which makes steam for electricity generation. A thermal storage system ensures that even more power can be generated by producing heat even after the sun goes down. (See "Storing Solar Energy," page 88.) This technology was first tried in Italy and France, but the United States was the first to apply it with two large multi-megawatt commercial power plants in California's Mojave Desert. Although these projects have ended, they sparked worldwide interest, particularly in Spain, Israel, and Australia, and several new "power tower" designs are being installed in California and throughout the Southwest United States.

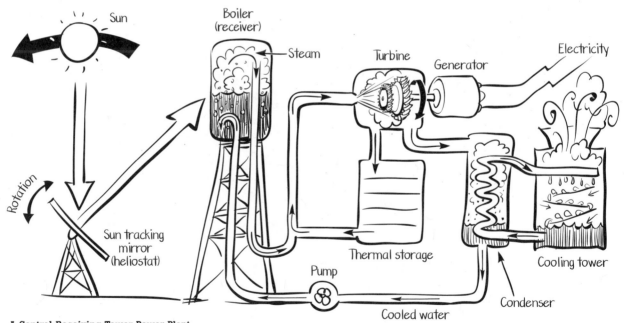

A Central Receiving Tower Power Plant
Though there is only one mirror shown in this diagram, in reality the tower is surrounded by many sun-tracking mirrors.

CONSIDERATIONS

- Solar energy systems have several attractive features. They are modular (units can be added as needed), make no noise, produce no pollution, and operate during the hours of highest daytime electrical demand.

- There have been two main challenges to the use of solar energy: the availability of sunlight "fuel" and the cost of the technology. Solar resources depend on time of day, the season, the cloud cover, and the location. Today, most of these factors can be addressed with various solar energy storage systems. While cost used to be a major barrier with solar technology, it is dramatically less so today. Costs have dropped in the past two decades, though solar systems are still expensive when compared with other renewables or with coal or natural gas.

- PV panels can be mounted on rooftops or even integrated right into the buildings as walls, skylights, sunshades, shingles and more. PV panels can also act as roofs over parking lots to provide shade and protection from the rain. These systems can take great advantage of otherwise wasted real estate space.

- The rooftops or land area needed to construct a big commercial PV facility is very large. This can present too great a financial challenge with today's technology. Also, large installations can be disruptive to certain sensitive desert habitats, though in most cases, once constructed, they are considered environmentally benign.

(continued)

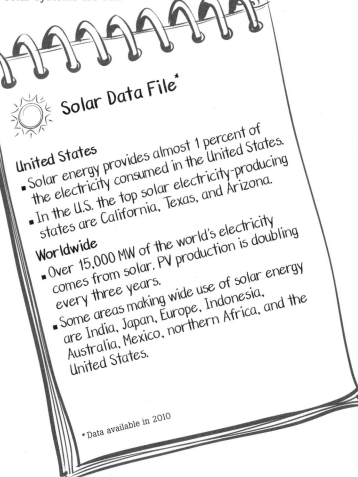

Solar Data File*

United States
- Solar energy provides almost 1 percent of the electricity consumed in the United States.
- In the U.S. the top solar electricity-producing states are California, Texas, and Arizona.

Worldwide
- Over 15,000 MW of the world's electricity comes from solar. PV production is doubling every three years.
- Some areas making wide use of solar energy are India, Japan, Europe, Indonesia, Australia, Mexico, northern Africa, and the United States.

*Data available in 2010

CONSIDERATIONS (continued)

- Large-scale solar power plants — and many other power plants that use renewable energy — tend to be located in remote locations, far from population centers or transmission lines. The challenge is to transport the energy from where it is produced to where it is used. One way would be to use solar (or other renewable) electricity to produce hydrogen from water through the process of electrolysis. The hydrogen could be shipped in containers or piped, just as natural gas is piped, to places where it could be used as a clean-burning fuel. (See "Producing Renewable Hydrogen," page 108.)

- Small Concentrating Solar Power (CSP) units do not take up much space and therefore can be placed in populated areas, especially industrial or commercial locations. These CSP units only work in the world's best solar resource regions.

- Manufacturing photovoltaic cells takes quite a bit of electricity. It takes two to four years to generate enough electricity from photovoltaic cells to compensate for the original electricity used to make them. The cells generally last 20 years or more.

- A solar thermal plant (CSP facility) can run reliably as a baseload power plant through the use of heat storage systems (see "Storing Solar Energy," page 88) or by supplementing solar with other fuels, like natural gas. CSP and PV produce electricity when the sun shines, which generally coincides with times of peak demand.

- Some renewable energy sources — notably solar — have benefited from recent government subsidy and rebate programs. These programs have given solar PV a big boost globally.

NO MONKEY BUSINESS

Solar energy is an attractive alternative for island nations, where geography makes the installation of transmission wires difficult and expensive. Stand-alone PV or solar dish engines could provide populations with electricity and reduce the need to cut down trees for firewood. For example, PV is being installed in many rural villages of the island nation of Indonesia, where it is hoped that orangutans and other threatened animals will benefit from the reduction of logging and its negative effects on their habitat.

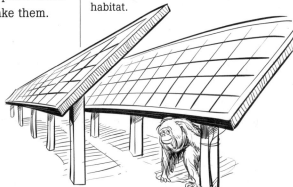

Creating Electricity from Solar Resources

Solar Troughs in Spain
This solar farm in Puertollano uses parabolic troughs — 352 collectors with 120,000 mirrors! — to focus the sun's energy on tubes of water, creating steam to drive turbine generators. *(Photos courtesy of Iberdrola Renovables, Spain)*

Dish-Stirling Systems
Prototypes of parabolic concentrators in Spain (above, left) track the sun to reflect its rays onto a solar heat exchanger, located at each concentrator. A Stirling engine and generator convert the heat into electricity. The photo on the right shows a different design that uses several concentrators to provide energy for a single engine. *(Photos courtesy of NREL, U.S. Department of Energy)*

Dual Purpose in Canada
These photovoltaic panels at the University of Calgary serve as both electricity generators and creative energy-saving sun shades.
(Photo courtesy of Conergy, Santa Fe, NM)

Solar Balloons
This invention is for off-grid uses, such as remote buildings or disaster sites. Filled with helium, a 10-ft. balloon could provide about a kilowatt of electricity. *(Photo courtesy of Dr. Joseph Cory, Geotectura Studio, and Dr. Pini Gurfil, Technion)*

Creating Electricity from Solar Resources

Good Use for the Nevada Desert
At Nellis Air Force Base 140 acres of previously unused land holds over 72,000 photovoltaic panels, built to generate 15 megawatts of electricity. *(Photo courtesy of U.S. Air Force)*

Solar Power Towers
Each 531-ft. solar power tower is the focus of 1,255 mirrored heliostats in Seville, Spain. Each heliostat, with a surface area of 1,291 square feet, reflects solar radiation onto the receiver, making steam to power a turbine generator. *(Photo courtesy of Abengoa Solar)*

Rocky Mountain Solar
Panels can be installed on any building — from a tiny cabin in a remote location to a skyscraper in an urban area. Mounting can be fixed (south-facing) or movable to track the sun. *(Photo courtesy of NREL, U.S. Department of Energy)*

Electricity Reaching Everyone
The photo on the left shows the *Digital Doorway*, a computer kiosk powered by weather-proof, durable solar panels. Units like these provide education and training to people in rural and disadvantaged areas of Uganda that have no electricity. The photo on the right shows a *Solar Suitcase*, a portable, user-friendly solar electric system that provides power for maternity hospitals and clinics in the developing world. The solar suitcase was first used in Nigeria and is now in nine countries. *(Photos courtesy of Sean Blaschke, UNICEF, and Laura Stachel, WE CARE Solar)*

Renewable Energy Source:
WIND

TERMS IN GLOSSARY

air current
anemometer
controller
fixed-speed wind turbine
jet stream
multi-megawatt turbine
nacelle
rotor
stand-alone wind turbine
terrain
variable-speed wind turbine
vertical axis turbine
wind farm

THE POWER OF WIND has been helping humans do work for centuries. As early as 5000 B.C. boats propelled by wind sailed along the Nile River. Windmills may have been used in China by 200 B.C., and by 900 A.D. large windmills were grinding grain on the plains of Persia. The windmill spread to England as early as 1100 A.D. and was a common sight throughout medieval Europe. In the 1800s the American West was settled with the help of thousands of water-pumping windmills.

The first wind turbines for generating electricity were designed in Europe around 1910. These soon appeared in the United States, bringing electricity to rural homes and farms. Beginning in the late 1930s, the widespread installation of power lines made these small wind turbines obsolete. However, this was not the wind turbine's last appearance "down on the farm." Today, the wind turbine is once again a familiar sight — on open plains, along mountain passes, even off of windy coastlines. Far advanced from their creaky windmill cousins, today's wind turbines are sleek and powerful contributors to today's electricity scene.

THE WIND RESOURCE

Wind patterns vary greatly from one place to the next. For example, one area in the middle of Ohio is consistently calm, while the winds off nearby Lake Erie can almost knock a person over.

Regional wind patterns are greatly affected by terrain and by air currents in the upper atmosphere. For example, upper-level winds (the jet stream) are a primary factor in the weather systems that bluster through the American Great Plains. The flat terrain in this area, offering no obstruction to the wind, also helps make this one of the windiest regions in the United States.

Experts who study wind patterns have developed a scale of 1 to 7 to classify wind power (wind speed, wind height, and other factors). Class 1 has the least power; Class 7, the highest. Wind turbines operate best in winds from Classes 3 through 7. There are many places in the U.S. where wind resources are Class 3 or above, including large parts of the Great Plains, the windy passes of the large mountain ranges, sections of both coasts, and portions of Alaska and Hawaii. Some wind experts believe that U.S. wind resources, if developed, could match total current U.S. electricity generation.

GENERATING ELECTRICITY FROM WIND RESOURCES

The basic machinery that converts wind power to electricity is called a wind turbine. The force of the wind spins blades attached to a hub that turns as the blades turn. Together, the blades and hub are called the rotor. The turning rotor spins a generator, producing electricity.

There is also a controller that starts and stops the rotation of the turbine blades. The generator, controller, and other equipment are found inside a covered housing (nacelle) directly behind the turbine blades. Outside, an anemometer measures wind speed and feeds this information to the controller.

Modern wind turbines are "variable speed." Their turning speed changes as wind speeds change, allowing them to withstand gusty gales and capture energy from both lower and higher wind speeds. (Earlier turbine designs, though still in use, are less efficient. They automatically shut off at high wind speeds.)

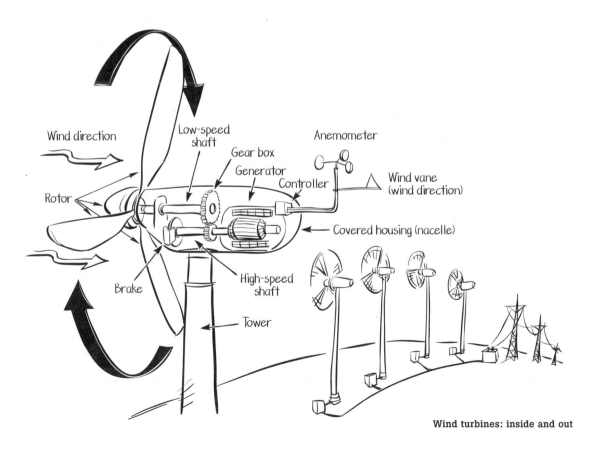

Wind turbines: inside and out

POWER SKETCH: Harvesting the Wind

Along Buffalo Ridge, where the Minnesota winds run high, midwestern farmers have found a new cash crop: the wind itself. They have harvested many benefits from leasing their land for the installation of electricity-generating wind turbines. Strong local winds generate electricity for at least 200,000 residents in the area. Some farmers report earning more than 10 times what they would make growing corn on the small area occupied by each wind turbine. Some enjoy the idea of producing non-polluting "green" electricity; others appreciate the boost to the local economy from increased jobs and even tourism. Most residents of Buffalo Ridge like wind farming so much that several communities hold autumn wind festivals, similar to the harvest festivals of the past.

The most common types of wind turbines have three fan-like blades that are usually placed at the top of a high tower. Vertical axis "eggbeater" designs and other innovatively shaped turbines are being developed today for use in household and urban settings.

Stand-Alone Wind Turbines

Remote Locations. Since a wind turbine is a complete "mini-power plant," one alone can be used for electricity with no connection to transmission lines. These stand-alone wind turbines are useful in rural and remote locations. They are ideal for village power in developing countries, though they need the back-up of a diesel or other type of generator or battery for continuous power needs.

A remote scientific research station at Black Island, Antarctica, uses stand-alone wind power for satellite communications and other systems. In spite of winds up to 175 mph (300 km/h) and -60°F (-51°C) winter temperatures, the turbines are remarkably reliable. In the extreme weather conditions of Alaska, several small wind turbines are providing power to remote villages in Kotzebue and Wales. This saves thousands of dollars in diesel fuel and greatly reduces pollution.

Areas with Transmission Lines. Stand-alone wind turbines are not only for remote areas. They can also be connected to electric distribution lines. Some people install a wind turbine in order to reduce their electricity bill and, frequently, to get some or all of their electricity from a clean renewable resource.

A good example is a large stand-alone turbine installed at a school in Spirit Lake, Iowa. The turbine powers the entire school, while educating students about the values of using renewable energy. Since the school doesn't have to buy very much electricity from its local utility, wind power saves the school around $25,000 a year in utility bills.

Wind Farms

Wind turbines are often grouped together into "wind farms" that send electricity to the local power grid. These energy farms can be completely compatible with other land uses, such as agriculture or livestock grazing.

Texas, Iowa, and California are the largest producers of wind energy in the United States. California, the state that pioneered the world's wind industry, now produces close to 2,800 MW of wind power. The state is host to several enormous wind farms in some of its windiest passes. At California's Altamont Pass, for example, thousands of wind turbines generate electricity for the nearby San Francisco Bay Area. Wind farms now dot Texas' windy ranch country, the largest of which is the Horse Hollow wind farm. Texas currently has over 9,000 MW of wind-generating capacity.

Wind turbines can now be seen on farms alongside crops and cows in windy rural areas of the Midwest. So far the two largest Midwest wind power producers are Iowa (over 3,000 MW) and Minnesota (over 1,800 MW).

REMINDER

W = watt
kW = kilowatt = 1,000 watts
MW = megawatt = 1,000 kilowatts

Using wind energy, 1 megawatt serves about 300 homes in the U.S.

Other states producing notable amounts of wind power include Washington, Oregon, Wyoming, Colorado, and New Mexico. The U.S. is now the world's largest producer of wind-generated electricity. Wind power capacity in the U.S. has increased by over 50 percent since 2008.

Wind energy is also a big industry in Europe. Germany is the world's second largest producer of wind power, followed by Spain. In Denmark and the Netherlands, modern wind turbines are seen along-side the old windmills for which this area is famous. Great Britain is installing many smaller wind farms of between 10 and 100 turbines each, both on agricultural land and off shore. Northern Africa, India, and a number of other areas are also investing in wind power. China is expected to emerge as a global leader in wind power in the next few years.

Wind Turbine Sizes

Wind turbines are usually divided into two categories: small-scale and large-scale.

Small-scale turbines generate less than 100 kW. These can be the principal power source in a remote location with some type of back-up power. Small-scale wind turbines can also be used to supplement existing sources of electrical energy, such as PV panels on buildings or electricity from the local utility. A "home-sized rotor" has blades that range from 8 to 25 feet (2 to 8 meters), meas-ured end to end. Found on towers reaching up to 120 feet (36.6 meters), most small-scale wind turbines can provide plenty of power for an average-size home, farm, or small business.

FOR THE BIRDS

One widely publicized concern is the issue of birds flying into spinning wind turbines. In fact, structures such as buildings, trans-mission lines, and communications towers pose more threats to birds than wind turbines. Nonetheless, no one wants to see any birds injured. A number of government agencies (including the U.S. Department of Energy) and many private companies are taking steps to reduce the poten-tial impact of wind power on our flying feathered friends. Researchers study sites for bird activity before wind turbines are placed. New smooth tubular towers (with no girders) discourage nest builders. With today's large multi-MW turbines, the distance between towers has increased, and their blades spin more slowly. Researchers are also trying out ways to make wind turbine blades more visible to birds.

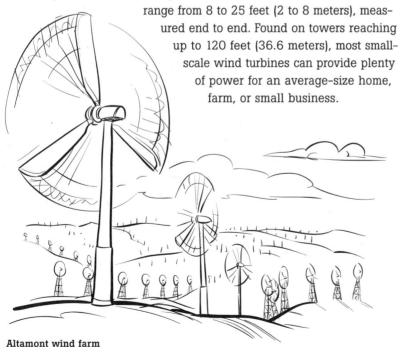

Altamont wind farm

Large-scale turbines range from 100 kW models to the newest, multi-megawatt machines. The most common large-scale turbines — often called "utility scale" turbines — each generate from 1 to 3 MW (and even 6 MW). The most common utility-scale turbine in the U.S. is 1.5 MW in size, but wind turbines as large as 10 MW are now being researched. These would be used exclusively for offshore sites.

Multi-megawatt turbines are enormous, with blades the length of a football field and towers twenty or more stories high. When installed offshore, these huge turbines take advantage of the sea's high winds. Great Britain, Germany, Denmark, and the Netherlands are already placing some of these mammoths off Europe's windiest shores. Great Britain has installed the world's largest offshore wind farm — 194 MW total — off its windy coast and is planning 18 more. The U.S. is looking at developing offshore wind farms along the Atlantic seaboard. For the West Coast, where the continental shelf drops off relatively close to shore, the industry is considering deepwater wind technologies.

WIND IN THE CITY

One interesting concept currently being explored (and not yet proven to be reliable) is the placement of wind turbines on buildings in urban settings. These little stand-alone turbines are being designed to harness the winds that rush through the human-built terrain of "hills and valleys" created by clusters of tall buildings.

CONSIDERATIONS

■ Today's sturdy wind turbines stand up to all kinds of weather. They are much more efficient than their earlier counterparts. Wind power is clean, since it provides electricity without having to burn fuels that create pollution.

■ The cost of producing electricity from wind power has gone down considerably in recent years, so that it doesn't cost much more than conventional fossil fuel technologies. Improvements in the equipment have made wind power very attractive as an energy source.

■ Wind velocity and direction are usually predictable within 24 hours and follow daily and seasonal patterns. But, of course, the wind doesn't blow all the time. (On average, wind turbines operate at about 30 percent of their full capacity.) While some wind energy can be stored in batteries, most on site, stand-alone systems also need some type of back-up. This is especially true in remote locations where people produce their own electricity. A common solution is a combination solar/wind system, along with batteries and perhaps a fueled generator.

■ Some people object to the noise of wind turbines, especially the older models. Newer turbines are usually no more audible than the wind itself. And in rural or remote areas, noise is not usually a significant issue. Because today's utility-scale turbines produce more power than earlier models, modern wind farms contain fewer turbines per acre. This reduces noise, visual impacts, and bird collisions.

■ In the United States, most electricity suppliers use wind power to supplement another energy source. It reduces the load on other plants operating at the same time, whether baseload or peaking.

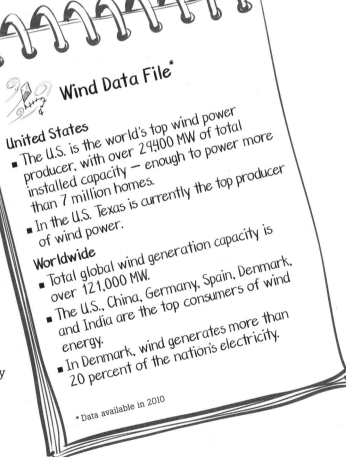

Wind Data File*

United States
■ The U.S. is the world's top wind power producer, with over 29,400 MW of total installed capacity — enough to power more than 7 million homes.
■ In the U.S. Texas is currently the top producer of wind power.

Worldwide
■ Total global wind generation capacity is over 121,000 MW.
■ The U.S., China, Germany, Spain, Denmark, and India are the top consumers of wind energy.
■ In Denmark, wind generates more than 20 percent of the nation's electricity.

*Data available in 2010

Creating Electricity from Wind Resources

Wind Energy for California Olive Oil

At 148 feet tall, this wind turbine north of San Francisco generates 225 kw of electricity — enough to power this ranch's olive oil processing plant plus all of the other ranch activities. *(Photo courtesy of McEvoy Ranch, Petaluma, CA)*

Wind Off Europe's Shore

Wind power makes increasing use of ocean expanses, where the wind blows harder and larger turbines can be installed. Many offshore wind farms are being developed in densely populated Europe, where there is limited space on land and relatively large offshore areas with shallow water. *(Photo courtesy of General Electric, Schenectady, NY)*

Vertical Axis Wind Turbines

These vertical-axis wind turbines are relatively new. They were developed for use in urban settings where traditional wind turbines might not be desirable or functional. If you look very hard, in the photo on the right, you can see a few of these turbines on the flat area between these buildings of Adobe Systems of San Jose, CA. *(Photos courtesy of Mariah Power, Reno, NV)*

Wind at Home and at Sea

This wind turbine model has unique curved blades for quiet operation. The turbine in the photo on the left generates about 50 percent of the power used by this home. In the photo on the right, a similar curved-blade turbine charges batteries for a sailboat in the West Indies. This small model also is suitable for RVs, remote cabins and offshore platforms. *(Photos courtesy of Southwest Windpower, Flagstaff, AZ)*

Creating Electricity from Wind Resources

Training in the Nacelle
Wind Technicians train at the GE Energy Learning Center in Schenectady, N.Y. The pictured hub is the tip of the nacelle; the big circles show where blades will go. *(Photo courtesy of Jill Gagnon and General Electric. Schenectady, NY)*

Maintenance
At the Shiloh Wind Power Plant in northern California, maintenance workers climb into the towers and up into the nacelle. *(Photo courtesy of Iberdrola Renewables, Radnor, PA)*

Multiple Uses of Land
Horses run freely at Shiloh, CA, among 100 turbines, showing the compatibility of wind power with ranching. *(Photo courtesy of Iberdrola Renewables, Radnor, PA)*

Wind Farms North and South
The photo on the left shows the unique black blades on wind turbines at Buffalo Ridge, MN. The blades' dark color helps shed frost and ice in winter. The wind farm shown on the right is near Palm Springs, CA. Here, wind power has turned unused desert land into a valuable wind resource. *(Buffalo Ridge photo courtesy of National Renewable Energy Laboratories, Golden, CO; Palm Springs photo courtesy of Iberdrola Renewables, Radnor, PA)*

The Renewable and Nonrenewable Resource

The Renewable and Nonrenewable Resource: HYDROGEN

TERMS IN GLOSSARY

anaerobic digestion

anode

cathode

compound

electrochemical

electrode

electrolysis

electrolyte

element

energy carrier

gasification

internal combustion
  engine

NASA

steam reforming

HYDROGEN IS ONE OF THE MOST ABUNDANT elements on Earth. Yet it wasn't until the 1700s that scientists first proved its existence, and it was later still that they recognized its value. Finally, by the mid-1800s, people were using hydrogen in "town gas," providing light and heat in cities across the United States and Europe. More recently, it has become useful in the production of ammonia, fertilizers, glass, refined metals, vitamins, cosmetics, cleaners, computers, and much more.

Hydrogen has launched many U.S. rockets into outer space. And hydrogen fuel cells, first used successfully in the 1960s, have been the main power source aboard all of NASA's space shuttles. Over the last 30 years, researchers have also been looking at other ways to use hydrogen as a fuel for everyday life.

Hydrogen: Renewable or Nonrenewable?

Hydrogen can be renewable or nonrenewable, depending on how it is produced. If it comes from a renewable resource (such as agricultural waste or water) and is produced using electricity from renewable energy, it is renewable. Otherwise, the hydrogen is considered nonrenewable. Most hydrogen produced today is nonrenewable.

THE HYDROGEN RESOURCE

On Earth, hydrogen is the third most common element, yet most of us aren't very familiar with it. This is because hydrogen doesn't occur naturally by itself. Instead, it is almost always found in combination with other elements.

Water, as we know, is a compound made of the elements hydrogen and oxygen — hence the formula H_2O. Hydrogen joins with carbon to make fossil fuels such as natural gas, coal, and petroleum. It is a main building block of life. Hydrogen is found in the molecules of all living things.

Water

Propane
(a fossil fuel)

Hydrogen: An Energy Carrier

In its natural state — in compounds with other elements — hydrogen's energy cannot be readily used. But hydrogen can *carry* the energy from those compounds to be used or stored, much as electricity carries energy from its source to its user. Hydrogen (like electricity) is therefore called an energy carrier.

Hydrogen Unbound

In order to use hydrogen we must separate it from the compounds in which it is bound. Hydrogen, once freed, is a colorless, combustible, carbon-free gas that carries a great deal of energy. Scientists have developed several different ways to produce hydrogen.

Producing *Renewable* Hydrogen

By Electrolysis. Electrolysis was first closely studied in the 1830s by English scientist Michael Faraday. In the electrolysis process, electricity is passed through water. The electrical charge causes the hydrogen and oxygen in the water molecule to split apart and turn into gases. An electrolyte, which may be a chemical or solid material, is often added to the water to help conduct electrons through it.

Water used in electrolysis is, of course, a renewable resource, but for the resulting hydrogen to be considered renewable, the electricity for this process must also have come from a renewable source. Any renewable method of generating electricity could be used.

HYDROGEN TO GO

One day our cars may operate using hydrogen gas. Some major car manufacturers have designed engines that burn hydrogen instead of gasoline. The engines of these cars are similar to those in the vehicles we drive today.

People like the idea of using these engines because burning hydrogen produces few polluting emissions. Researchers are now working on ways to store hydrogen aboard a vehicle, along with ways to make hydrogen "filling stations" widely available. Some companies are developing vehicles that use hydrogen without combustion. (See "Fuel Cells," page 110.)

POWER SKETCH: Hydrogen Fuel Cells Keep Aquarium Bubbling

A unique energy system, the Schatz Solar Hydrogen Project, helps keep fish alive in the marine lab aquarium at Humboldt State University in northern California. Here, solar panels mounted on the roof produce electricity that is used for two purposes: to drive the aquarium's aerator (which adds oxygen to the water for the fish) and, at the same time, to produce hydrogen gas by electrolysis. The hydrogen gas is stored and then, whenever the sun doesn't shine, the hydrogen is used in a fuel cell to provide electricity for the aquarium's aerator. This remarkable renewable energy system has been running day and night since 1994.

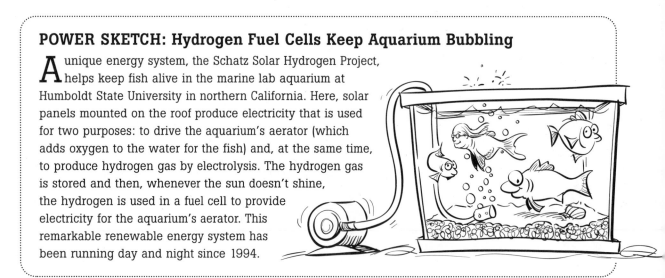

In the future, electrolysis systems might be installed at renewable energy power plants. Some or all of the electricity could be used for electrolysis, producing hydrogen gas that could be transported and used for other purposes. Electricity could be produced for customer use when needed, then shifted to use for hydrogen production at times of lower electricity demand.

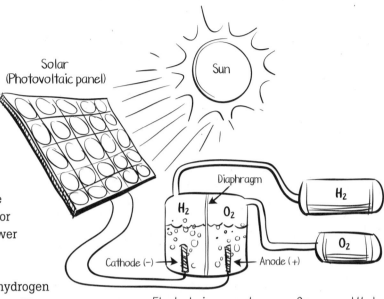

Producing renewable hydrogen using electrolysis

Using Biomass. Biomass gives off hydrogen gas when it's heated in a certain way. Plant material (such as tree trimmings or specific crops) or organic waste can be used in this process, called gasification. Gasification is a thermal process that converts organic material into hydrogen, carbon monoxide, carbon dioxide, and small amounts of other gases. This mixture is frequently called synthesis gas, or syngas, because it can also be used to produce other chemicals.

Using Landfill Gas. When organic material begins to break down in our landfills, it gives off gases such as methane. Hydrogen can be produced from this methane gas and then used to generate electricity with a fuel cell. Hydrogen production from methane gas does give off carbon dioxide (as is also true with producing hydrogen from fossil fuels). However, since methane is a more potent greenhouse gas than carbon dioxide, using it to produce hydrogen is still considered preferable to allowing it into the atmosphere.

Using Biological Organisms. Some micro-organisms produce methane and other gases when they are caused to digest under special conditions; this process is called *anaerobic digestion*. Hydrogen can be extracted from the resulting methane gas. (See "Munching Microbes," page 40.) Anaerobic digesters can be used for a variety of purposes, including cleaning waste water and converting industrial waste.

Producing *Nonrenewable* Hydrogen

By Electrolysis. Hydrogen can be produced by electrolysis (described on page 108) using electricity from either renewable or nonrenewable resources. If the electricity comes from a fossil fuel or nuclear plant, then the resulting hydrogen is nonrenewable.

By "Steam Reforming." Another method of producing hydrogen involves using a fossil fuel, such as natural gas, and steam to produce hydrogen and by-products, in a process called *steam reforming.** This method involves boiling equal amounts of natural gas and water to produce hydrogen and byproducts. The high-temperature steam separates hydrogen from the carbon atoms in the fossil fuel. Steam reforming is the process most commonly used today to produce hydrogen.

By Gasification. Gasification is described on pages 40 and 109 as a means of producing gas from biomass. A similar process can be used with a fossil fuel to produce hydrogen.

GENERATING ELECTRICITY FROM HYDROGEN

Fuel Cells

Fuel cells were once thought to be a "far out" technology suitable only for use aboard space shuttles. However, fuel cell technology is advancing rapidly. Fuel cells are already popping up in many aspects of everyday life, including generating electricity, powering vehicles, and operating small electrical devices. Even fuel-cell powered toys are now available.

Most of us associate the word "fuel" with burning something for its energy. However, in spite of the name, nothing is burned in fuel cells. Instead, fuel cells produce electricity using a method that is actually the reverse of electrolysis. Hydrogen (the "fuel") and oxygen are combined (rather than separated) through an electrochemical process that produces electricity, heat, and water.

Some types of fuel cells can use liquid or gas hydrocarbons directly, although most can use them only if the hydrocarbons are first converted into hydrogen.

The steam reforming method can also be accomplished using certain geothermal resources. When geothermal steam is used, the resulting hydrogen is renewable.

DISPELLING A MYTH

On May 6, 1937, a flash fire engulfed *The Hindenburg*, a luxury zeppelin aircraft filled with hydrogen gas. Dozens of people were killed. As the flaming airship plunged to the ground, newsreel cameras captured the disaster. The film footage caught the world's attention, and for decades hydrogen gas took the rap for causing the fire. Recently, hydrogen expert (and former NASA researcher) Addison Bain proved conclusively that hydrogen was not to blame. Rather, the fire was caused by the design and highly flammable fabric covering the craft, working in deadly combination with electric sparks from a developing thunderstorm. Bain's findings were confirmed by eye-witnesses who described the fire as a bright, fireworks-like display of color. Hydrogen would have burned with a colorless flame.

Fuel cells produce no polluting emissions. And, if the hydrogen used is produced with renewable methods, then the fuel cell is also considered renewable.

Because fuel cells are so clean, some states already provide financial incentives, or even exemption from permitting requirements, for fuel cell projects. Fuel cells range from very small units to those that produce more than one megawatt. Because they are modular, extra units can be added when more power is needed.

Hydrogen as a Combustible Fuel

Hydrogen can also be used as a combustible fuel, in either a liquid or gaseous state. Hydrogen burns completely with very few pollutants and has a high energy content.

Sometimes hydrogen is added to natural gas at traditional power plants, making them work more efficiently and helping to reduce pollutants. It is possible that these power plants could be remodeled to run solely on hydrogen gas. If the hydrogen gas came from a renewable source, then these updated power plants would both supply renewable power and be easy on the environment.

Most of the current attention paid to combustible hydrogen fuel is for use with turbines, which would be used in a smart grid. (See "Smart Grids," page 147.)

> **REMINDER**
>
> **W** = watt
> **kW** = kilowatt = 1,000 watts
> **MW** = megawatt = 1,000 kilowatts
>
> 1 megawatt can serve about 1,000 homes in the United States.

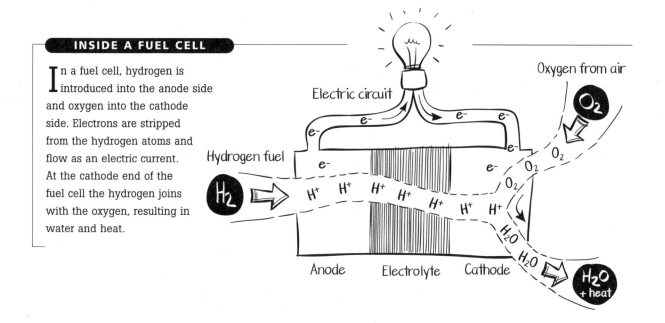

INSIDE A FUEL CELL

In a fuel cell, hydrogen is introduced into the anode side and oxygen into the cathode side. Electrons are stripped from the hydrogen atoms and flow as an electric current. At the cathode end of the fuel cell the hydrogen joins with the oxygen, resulting in water and heat.

Hydrogen at Work in the U.S.

Around the country, many cities, utilities, hospitals, and industrial facilities are exploring the potential for fuel cells in electricity generation. Fuel cell power projects have already been installed in almost every state and new applications are being tried out. One example is a fuel cell that runs on hydrogen converted from methane, which was installed in a coal mine in Ohio. This fuel cell reduces dangerous coal mine methane emissions, while it provides electricity for miners. In Texas, a chemical manufacturer is using the hydrogen it produces onsite — which was once considered a waste — in a fuel cell that provides power for the factory. Below are some other examples of cool projects happening around the U.S.

A large fuel cell installation is providing power for a training school in Connecticut. At a wastewater treatment facility in Portland, Oregon, fuel cells using hydrogen produced from waste gas provide back-up power for plant operations. In New York, the Central Park Police Station runs on fuel cell power. The U.S. Department of Defense has already installed several stationary fuel cell power stations at a number of military bases.

Some major grocery chains are now using fuel cell forklifts, and telephone companies are using fuel cells to supply back-up power for cell phone towers and switching stations. A microwave relay station mounted atop a fire watchtower in Redwood National Park in northern California runs on both solar PV and fuel cell systems. A school in Santa Cruz, California, has a portable fuel cell unit that fits in a small suitcase. A renewable-energy teaching tool for the students, this system can run an ice cream maker, a blender, or even a computer.

The California Fuel Cell Partnership, founded over 10 years ago, links dozens of private and government groups to test fuel cell cars and to encourage building of hydrogen fueling stations. Hundreds of fuel cell cars have operated on California highways since 1999.

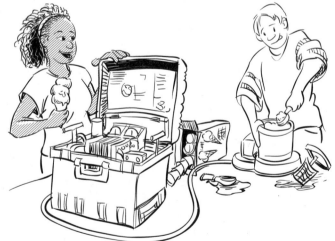

Portable fuel cell unit providing electricity to an ice cream maker

Image adapted with permission of the Schatz Energy Research Center

Hydrogen at Work Around the World

Iceland, already a leader in the use of hydropower and geothermal energy, is promoting the use of hydrogen to displace the 30 percent of its energy that comes from imported oil. Doing so would make Iceland completely energy self-sufficient. India is using fuel cells to meet exploding demand for cell phones. Korea has plans to be a huge supplier of fuel cells and hopes to create over half a million jobs in this industry. Many European countries, along with India, Japan, Korea, China, Australia, and others, are also pursuing the use of hydrogen fuel cells.

CONSIDERATIONS

- Hydrogen is a transportable fuel. This is important because many renewable energy resources are not transportable. In remote and unpopulated areas, far from transmission lines, local geothermal, ocean, solar, and wind resources can be used to produce electricity. Then the electricity can be used on-site to make renewable hydrogen that can be stored to provide electricity when power is not available and/or transported for use elsewhere as a combustible fuel or in fuel cells.

- If hydrogen escapes from its container, it rapidly disperses into the air rather than puddling on the ground the way heavier-than-air fuels, such as gasoline, tend to do. However, hydrogen burns easily and invisibly, so care needs to be taken when handling it, especially if it escapes and collects in a contained space. (Hydrogen is explosive if exposed to oxygen.) With proper precautions, hydrogen is thought by some to be just as safe as gasoline. Currently, engineers are perfecting systems to contain and transport hydrogen safely and economically because it is considered an important fuel for our future.

- Hydrogen burns cleanly, though it does produce some emissions when burned. Used in a fuel cell, the only by-products are heat and water.

- Hydrogen has about three times the energy of gasoline by weight, yet only one third as much energy by volume. Hydrogen storage is therefore an issue that could limit future applications. Scientists are working hard on new storage options.

(continued)

CONSIDERATIONS (continued

■ Currently, much of our hydrogen gas comes from processes that use fossil fuels. If we continue producing hydrogen in these ways, we will probably have the same concerns about hydrogen production that we now have about fossil fuel use — i.e., energy insecurity, depletion, and pollution. (See pages 122–123, "Considerations," and all of Chapter 4, "Energy, Health, and the Environment.")

■ The production of hydrogen, especially with renewable methods such as electrolysis, is still quite expensive. It is hoped that costs will come down as the technology for producing renewable hydrogen is perfected through research and experience.

■ Some people think that hydrogen will replace fossil fuels as our primary source of energy. Conversion to a "hydrogen economy" will require new technology and distribution networks that will take years to develop. In the U.S. enthusiasm for hydrogen as a power option has diminished over the last decade. However, according to industry experts, Europe, Asia and the Middle East, still view hydrogen as a strong competitor as a fuel option and also as a storage medium for renewable energy.

■ Within the fuel cell industry, some consider hydrogen to be renewable if it is produced from a renewable resource, even if by a process using nonrenewable electricity.

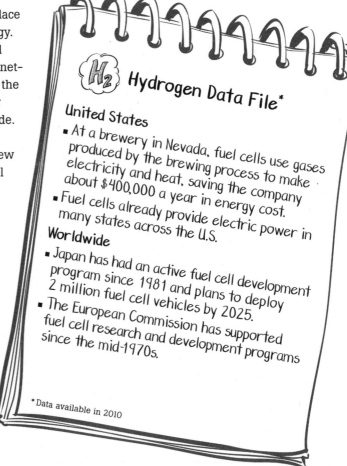

H₂ Hydrogen Data File*

United States
■ At a brewery in Nevada, fuel cells use gases produced by the brewing process to make electricity and heat, saving the company about $400,000 a year in energy cost.
■ Fuel cells already provide electric power in many states across the U.S.

Worldwide
■ Japan has had an active fuel cell development program since 1981 and plans to deploy 2 million fuel cell vehicles by 2025.
■ The European Commission has supported fuel cell research and development programs since the mid-1970s.

*Data available in 2010

Nonrenewable Energy Sources

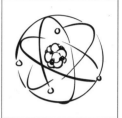

Nonrenewable Energy Source: FOSSIL FUELS

TERMS IN GLOSSARY

acid rain
carbon-based compound
combined cycle power plant
crude oil
gas turbine
global climate change
greenhouse gas
hydrocarbon
liquefied natural gas (LNG)
oil rig
oil refinery
scrubber
synthetic

FOSSIL FUELS — COAL, OIL, AND NATURAL GAS — have been highly prized energy sources for centuries. Mining for coal may have first occurred in China as far back as 200 B.C. By 200 A.D. the Romans made wide use of the coal resources they found in the British Isles. In the 1100s, oil wells were being drilled in Europe and along the west coast of the Caspian Sea. It was the Industrial Revolution, however, that launched the widespread use of fossil fuels to power factories and transportation systems. Electricity was first produced using coal in the 1880s. Since that time, fossil fuels have been the dominant source of energy for electrical production, transportation, and industry in the United States and around the world.

THE FOSSIL FUEL RESOURCE

All fossil fuels were formed from plants and animals that lived millions of years ago — long before the days of the dinosaurs (hence, the phrase "fossil" fuels). When these plants and animals died, their remains decomposed and were eventually buried under tons of soil and rock. Subjected to heat and pressure over time, this organic matter eventually formed coal (a solid), oil (a liquid), and natural gas (a vapor). These three different fossil fuel types resulted from variations in geologic conditions over time.

Fossil fuels are nonrenewable resources. They formed very long ago, when much of the earth was covered with swamps and the climate was extremely warm. These conditions were perfect for many living things, including huge ferns, trees, and other plants. The swamps and seas were teeming with algae and other small organisms. These lush conditions are not nearly as widespread today. Small amounts of fossil fuels may still be forming, but not in significant quantities. And, they will not form in a useful amount of time.

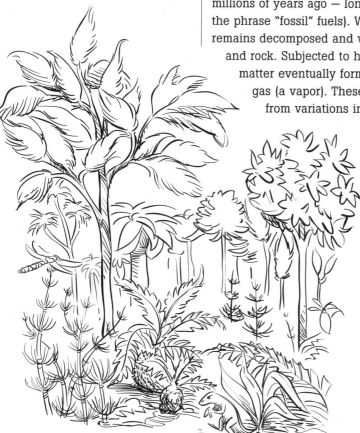

Plants and animals of long ago formed the fossil fuels we use today.

Living things are carbon-based, so all fossil fuels are made of molecules that contain carbon. They also contain hydrogen, giving rise to the name "hydrocarbons." Hydrocarbons burn easily. They are a reliable source of heat energy and are convenient to transport.

When fossil fuels are burned, carbon combines with oxygen, resulting in emissions of carbon dioxide gas. Fossil fuels contain other substances in addition to hydrocarbons. Sulfur, nitrogen, mercury and other impurities are found in varying amounts in each fossil fuel. When burned, these recombine with other materials and form air pollutants.

Coal

Coal is a solid hydrocarbon that we excavate from underground, just as we mine for minerals. One age-old method is to mine coal from tunnels dug deep underground. The other, and more recent, method is called surface- or strip-mining. Here, deposits within about 200 feet of the surface are exposed by removing the overlaying rock and soil. Once topside, coal is easy to transport, usually in large containers aboard ships or on trains.

There are abundant supplies of coal in the United States, with coal deposits in states across the continent. The top coal-producing states are Wyoming, West Virginia, Illinois, Montana, and Pennsylvania. Outside the U.S., China, Australia, India, and South Africa produce the most coal.

Oil

Oil, also known as petroleum or crude oil, is a thick black liquid hydrocarbon found in reservoirs hundreds to thousands of feet below the surface. We extract it by drilling wells deep into the underground rock and then inserting pipes. Natural pressure can bring the oil shooting to the surface when wells are new; but, in most cases, pumps are needed to bring the oil to the surface. These oil field pumping units are common sights on land and at sea (on offshore platforms) in oil-producing areas.

Once captured, crude oil is taken to refineries and processed into various products. These include gasoline, diesel, aviation fuel, home heating oil, asphalt, and oil burned for electrical power. Oil products are sent from refineries through pipelines directly to consumers, or are delivered in large tanks aboard trains, trucks, or tanker ships.

MAKING AMERICA GO

Though less commonly used for producing electricity than coal, oil is still the most widely used fossil fuel. Why? Because for decades it has been refined into gasoline, diesel, and aviation fuel to power our cars, trains, trucks, and planes. It is also used extensively for heating homes and businesses, for industrial process heat, and to make fertilizers, machinery lubricants, medicines, and many types of plastics.

A pumping rig is used to bring up crude oil.

Of the world's top producers of crude oil, Saudi Arabia is first, the United States is second, and Russia is third. In the U.S., oil is produced in Texas, Alaska, California, and 28 other states. Production in the United States has already begun its decline. Over half of the oil used in the U.S. today is imported, and experts expect Africa to supply much of the world's oil in the future.

Natural Gas

Natural gas (mostly methane) is a hydrocarbon vapor that occurs naturally underground. (It is not the same thing as the liquid gasoline that we use in our vehicles, though we do call this "gas" for short.) Natural gas is piped to the surface through wells drilled into the underground rock.

Natural gas can be processed into propane and other types of fuels. All natural gas fuels are highly flammable. Natural gas is odorless, so for safety it is mixed with a chemical to give it a noticeable smell before it is sent to consumers. Huge networks of pipelines deliver most natural gas directly to homes, factories, and power plants. Natural gas can be stored and shipped in pressurized containers. It can also be condensed to a liquid. This liquefied natural gas (also known as LNG) can be transported and re-vaporized for later use. Natural gas can also be used to produce nonrenewable hydrogen.

Russia, the United States, and Canada are currently the world's top producers of natural gas. As natural gas production in the U.S. and Canada continues to decline, more and more natural gas is likely to be imported as liquefied natural gas.

GENERATING ELECTRICITY WITH FOSSIL FUEL RESOURCES

Most electricity in the United States is produced in "conventional" fossil fuel power plants. A fuel is burned to boil water to make steam. The force of steam is what drives the turbine generator. (See "How a Steam-Driven Power Plant Works," page 29.) While some plants burn petroleum or, more frequently, natural gas, the fuel most used for electricity generation in the U.S. has been, and still is, coal.

> **REMINDER**
>
> **W** = watt
> **kW** = kilowatt = 1,000 watts
> **MW** = megawatt = 1,000 kilowatts
>
> 1 megawatt can serve about 1,000 homes in the United States.

Coal-fired Power Plants

A 1,000 MW coal-fired power plant burns about 10,000 tons of coal a day, providing electricity to about one million people. Sometimes a coal-powered plant is located right at a coal mine. Other times the coal arrives in trains that go back and forth non-stop between the mine and the power plant. The coal is usually processed into pulverized fine particles that are burned to create the steam needed to power the turbine generators.

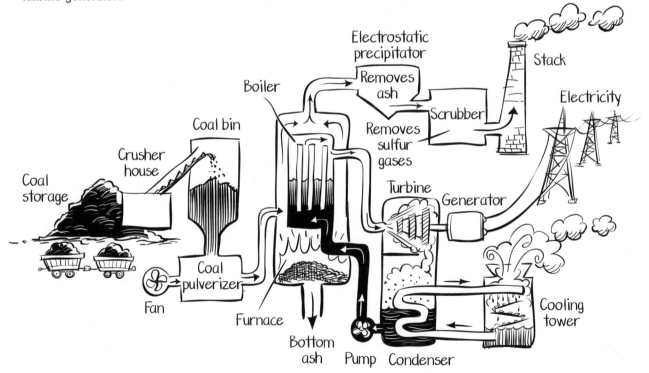

A conventional coal power plant

Currently we are developing methods to "clean" coal. These methods reduce the amounts of some impurities, such as sulfur and nitrogen, before the coal is burned. This process lessens the need to remove by-products after it is burned. We have also developed a technology to "cook" coal using gasification. This produces a cleaner-burning, synthetic (artificially made) gas. Coal gasification can also produce hydrogen.

Gas Turbines

Natural gas was originally used mostly for heating and in industrial processes. But now it is also used for generating electricity and was, in fact, the preferred fuel for new power plants in the 1990s. The first natural gas power plants were conventional steam-driven combustion turbines. Today's gas-powered plants use a turbine based on jet aircraft engine design. A mixture of compressed natural gas and high-pressure air is burned in a continuous fiery explosion. The hot exhaust from this combustion reaction is what drives the turbines. This method is more efficient and cleaner burning than the steam-driven turbines.

POWER SKETCH: Clean Spin on an Old Design

Recent advances in the design of gas turbines mean much greater efficiency and far less pollution. One of these high-tech turbines is larger than the biggest locomotive. It uses the same system as today's gas turbine models: the exhaust resulting from the explosive combustion of compressed natural gas is used to spin the turbine. What makes this turbine even better than earlier designs is its exceptional energy efficiency and greatly reduced air emissions. Planned to work in "combined cycle" power plants (see next page), these sophisticated turbines are being applauded as a cleaner way to produce electricity when using fossil fuels.

"Combined Cycle" Power Plants

Since about 1985, most new natural gas power plants have been "combined cycle" plants, in which two turbine types work together to produce electricity. First the gas turbine produces electricity using the hot exhaust from the combustion reaction as described on page 121. Then that same exhaust — still extremely hot — is used to boil water to produce steam in a conventional boiler. The steam spins a second turbine that generates even more electricity. Since combined cycle systems produce extra electricity by using what would otherwise be wasted heat, they are exceptionally energy efficient.

Size of Fossil Fuel Power Plants

Coal plants and natural gas power plants are typically large, generating 300 to 1000 MW of electricity or more. But natural gas plants come in all sizes. A university in Florida uses a 42 MW plant to produce electricity, hot water, and heat for its buildings. Often hospitals, or other places that must continue to operate at all times, have back-up diesel or natural gas turbines in case the power goes out. Natural gas "microturbines" of 25 to 500 kW can also be used in small businesses and as standby or peaking power.

CONSIDERATIONS

- Coal, oil and natural gas are high-energy fuels that are easy to transport. They can also be converted to other forms like propane.

- Fossil fuels have the advantage of a long history. Technology using fossil fuels has been refined over time, so their use is convenient and familiar.

- Forecasts differ for how long world oil and natural gas supplies will last at projected rates of consumption. There seems to be agreement that it will be only decades, not centuries. Many experts suggest oil supplies have already peaked.

- As oil and natural gas production decline worldwide (as it already has in the U.S.), prices will rise due to shortages or fear of shortages.

- Fossil fuels create air pollution when burned. In the United States, regulations require that most fossil-fuel power plants equip their smokestacks with "scrubbers" that trap some of these pollutants. Enforcement of regulations tends to vary with the political climate. (See also Chapter 4, "Energy, Health, and the Environment," pages 133-142, and Chapter 5, "Energy Management Strategies and Energy Policy," pages 143-155.)

- Natural gas has fewer impurities than coal or petroleum and burns cleaner than other fossil fuels.

- Most of the power plants built in the U.S. in the last 20 years have been natural gas plants. As U.S., Canadian, and Mexican natural gas supplies decline, the U.S. will need to import more and more liquefied natural gas in LNG vessels from distant ports.

- It takes energy to liquify and transport natural gas, which must be cooled to -260°F (-162°C) and be kept at that temperature. Then, at its destination port, it takes still more energy to return it to a vapor and pressurize it for delivery through pipelines to customers.

- Although grateful for worldwide fuel sources, many Americans have concerns about importing them: supply and cost impact the economy, and foreign dependence impacts national security.

- Every industry has accidents, and, while they are not common, some have greater consequences than others. For example, the 1982 Valdez oil tanker accident in Alaska created an oil spill that covered over 1,000 miles of shoreline. The oil killed many birds, fish, and other animals. It greatly disrupted the natural habitat and the local fishing economy.

(continued)

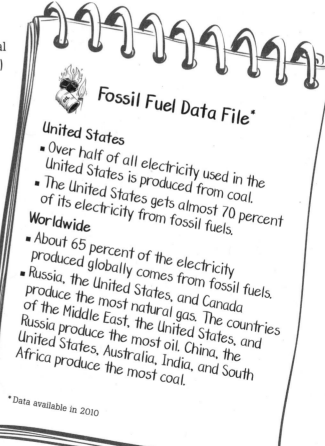

Fossil Fuel Data File*

United States
- Over half of all electricity used in the United States is produced from coal.
- The United States gets almost 70 percent of its electricity from fossil fuels.

Worldwide
- About 65 percent of the electricity produced globally comes from fossil fuels.
- Russia, the United States, and Canada produce the most natural gas. The countries of the Middle East, the United States, and Russia produce the most oil. China, the United States, Australia, India, and South Africa produce the most coal.

*Data available in 2010

CONSIDERATIONS (continued)

- Mining coal often causes serious disturbances to the surface habitat of an area. For example, to uncover the coal deposits at some surface coal mines, hilltops are scraped off, and the plants, soil, and rocks are pushed into the valleys and streams below. And with tunnelling, holes usually remain after a mine is abandoned. If soils are set aside and replaced, an agricultural area can usually be "reclaimed" and returned to farmland once the coal has been removed. However, any natural area disrupted by mining activity will in all likelihood never be the same.

- Coal and oil power plants are usually baseload facilities. Natural gas power plants can be operated as baseload or peaking plants, and small gas turbines are often used as emergency backup. Diesel plants are used for peaking or emergency standby power.

Nonrenewable Energy Source: NUCLEAR

TERMS IN GLOSSARY

chain reaction
containment vessel
control rod
fissionable
fuel rod
nuclear fission
nuclear fusion
nuclear reactor
plutonium
radioactive
reactive
reactor core
spent fuel
subatomic particle
thorium
uranium

This nuclear power plant uses ocean water for cooling; it does not need traditional cooling towers.

THE ATOMIC AGE WAS BORN in 1939 when physicists burst apart the nucleus of a uranium atom, releasing a tremendous amount of energy as heat and light. They called this reaction nuclear fission. (Fission means "to split.")

Nuclear fission's first job was to make atomic bombs during World War II (in the 1940s). However, we soon learned how to control the energy from nuclear fission so we could use it to produce electricity. Today, nuclear energy is used widely for electricity generation. It is also used to power Navy submarines and some aircraft carriers.

THE NUCLEAR RESOURCE

Nuclear energy is the energy trapped inside atoms, those tiny particles from which all matter is made.

The Energy of Atoms and Molecules

In nature, atoms are bonded together into molecules, which in turn are bonded into various types of matter. It takes a great deal of energy to hold these molecules together.

Every atom is made up of even tinier "subatomic" particles, including the protons and neutrons in the atom's nucleus (central part). The energy that holds these subatomic nuclear particles together is significantly greater than the energy that holds molecules together.

POWER SKETCH: A Natural Nuclear Reactor

Nuclear power plants depend on fissionable materials, which include radioactive elements. These materials will release the energy bound in their atoms in a nuclear chain reaction. In most cases, the radioactive element used is uranium. Uranium is so reactive that it will, under very special circumstances, produce its own atomic reaction without any human help. At the Oklo mine in the West African country of Gabon, a deposit of "spent" uranium was found deep underground. This uranium had at one time spontaneously become a natural "nuclear reactor." Millions of years ago, it began its own self-sustaining chain reaction that lasted about 500,000 years!

Making nuclear energy can be roughly compared to burning wood. When we burn wood, we produce energy by breaking the electron bonds between atoms and between molecules. If we stand beside a blazing bonfire we feel the energy of this chemical reaction as heat and see the energy as light. Similarly, when we produce a nuclear reaction, we break the bonds between protons and neutrons within the nucleus of each atom, releasing enormous amounts of energy — considerably more than our bonfire.

Uranium Nucleus is "Easy" to Split

Most of the elements found on Earth have stable nuclei (plural of nucleus). This means they don't split apart easily. But some elements, such as uranium, have unstable nuclei, which causes these elements to give off small particles (to "radiate"). One type of uranium, Uranium 235 (U-235) is especially unstable.*

Uranium: Fuel for Nuclear Power

Uranium is very hard and dense. That is, it has a lot of mass per given volume. Whereas one gallon of milk weighs about 8 pounds, one gallon of uranium weighs 150 pounds.

Uranium is found in many parts of the world, including the United States. We dig uranium-bearing rock (ore) from the ground just as we mine other minerals. There is a limited supply — though scarcity is less of an issue than it is for fossil fuels, since uranium is used in much smaller quantities. Uranium is, nevertheless, a nonrenewable resource.

Elements other than uranium, notably plutonium and thorium, can also be used for nuclear fission. In most parts of the world plutonium is only used in weapons and not for the production of electricity. Thorium has been used successfully in experimental reactors. It is estimated to be three or four times more abundant than uranium, though its commercial practicality has not been proven.

GENERATING ELECTRICITY WITH NUCLEAR ENERGY

A Nuclear Chain Reaction

In a nuclear power plant the process of nuclear fission — splitting uranium nuclei — is accelerated. Controlled for safety, the process produces enough heat for steam to power a turbine generator.

For nuclear fission to occur, high-energy subatomic particles, neutrons, are caused to bombard the uranium atom's nucleus, breaking it apart. When the nucleus splits, it releases heat and light as well as neutrons. These particles strike other uranium atoms, splitting those as well. These, in turn, strike and split other atoms, and so on, producing a nuclear chain reaction.

The amount of energy produced by splitting one uranium nucleus isn't much. However, because uranium is so dense, one pound of uranium has billions and billions of nuclei. Once we start a chain reaction, a LOT of energy is released. In a nuclear power plant, this reaction is carefully controlled to allow just the right release of heat energy needed to produce electricity.

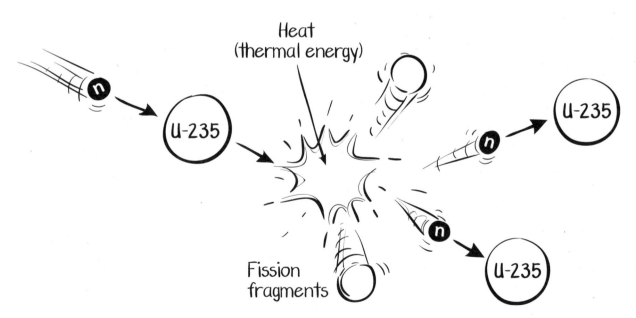

Starting a nuclear chain reaction

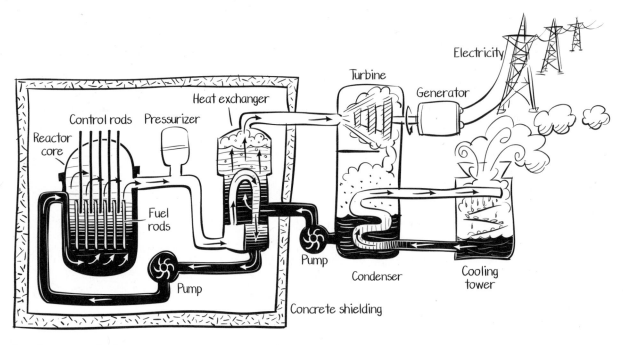

A nuclear power plant

Preparing the Nuclear Fuel

At a nuclear fuel processing plant, the natural uranium is "enriched" — a mechanical process that increases the U-235 concentration by about 4 percent to make the uranium more useful.* The enriched uranium is formed into pellets the size of the tip of your little finger. (Each pellet contains the energy equivalent of a pick-up truck full of coal, 150 gallons of oil, or a house-sized container of natural gas.) The pellets are loaded into long metal fuel rods.

Inside a Nuclear Power Plant

Many fuel rods are placed into a reactor core, interspersed with moveable control rods holding a material that absorbs neutrons. Pushed in or out of the core, the control rods govern the size of the reaction (and, therefore, the amount of energy produced).

*When used in nuclear weapons, the U-235 concentration must be enriched by 85-90 percent. The much lower concentration used in a nuclear power plant is calculated to avoid a sustained explosive reaction.

The fission reaction is started with high-energy neutrons from uranium inserted into the reactor. These neutrons bombard the pellets, splitting some of the uranium nuclei. This releases more neutrons, causing the chain reaction that produces heat and light.

A liquid or gas flows past the fuel rods in the reactor core, carrying some of the heat to a heat exchanger. (See "Heat Exchangers," page 53.) In the heat exchanger, this heat is transferred to water that boils and makes steam to spin the turbines. The rest of the system works like a traditional steam-driven power plant.

Nuclear Power Safety and Nuclear Waste

Uranium requires caution through many stages. It is less hazardous in its natural state than later; but danger of inhaling or swallowing exists while it is mined and milled, and dangers increase as it is enriched (made more radioactive), fashioned into pellets, and stored for use.

The amount of waste produced from nuclear reactors is small in comparison to the energy produced. The waste from one person's lifetime use of nuclear power has been estimated to occupy the space of a soda can, whereas waste from coal power over that time can be 68 tons of solids and 77 tons of carbon dioxide.

Nuclear waste, even in small doses, is lethal and remains hazardous for tens of thousands of years. Though quantities created at a given time are small, they build up over time, so that 72 storage facilities in the U.S. hold an estimated 50,000 tons of used (spent) fuel.

Spent fuel is still radioactive. It must be handled carefully and stored safely in a secure place. ("Secure place" is a matter of controversy; see "Considerations," pages 130-131.) Radioactivity decreases over time; some estimate a 99 percent drop in 40 years. But decrease does not mean elimination; the waste remains dangerous for many generations.

Nuclear Fuel Recycling

A portion of the spent fuel is removed periodically from the reactor, about every 18 months; each nuclear pellet may last 5 years. After removal, the fuel still retains most of its energy. If recycled (reprocessed), it could be reused in a modern, efficient plant. Some say present supplies could fire our nuclear plants for 40 years, with reprocessing, during which the volume of waste would shrink. Some countries reprocess their used fuel, but not the U.S.

NUCLEAR FUSION

Nuclear *fusion* (as opposed to nuclear *fission*) is another form of atomic energy. Nuclear fusion means several smaller nuclei are "fused" to form a larger nucleus. When this happens, a huge amount of energy is given off as heat and light. This process is what makes our sun produce heat and light, making it a natural nuclear reactor. Nuclear fusion is very appealing as an energy source, because it uses less fuel and creates less radioactive material than nuclear fission. However, scientists have not yet learned how to control fusion reactions that produce usable energy. So despite 50 years of effort, engineers haven't yet found a practical way to use a fusion reaction.

One reason nuclear fuel is not recycled in the U.S. is because a reprocessing byproduct is plutonium, usable in weapons. High cost of reprocessing is another deterrent. Other countries — notably Russia, France, and Japan — have government-supported reprocessing facilities.

Nuclear Power Around the World

In some parts of the world, mainly in industrialized countries such as the U.S., nuclear plants have not been built in several years, although support has grown recently for building more. In Asia, particularly China, South Korea, and India, electricity production from nuclear power plants is expected to increase.

CONSIDERATIONS

- Nuclear power is among the most controversial issues of our age. By wide agreement, great care is necessary in producing, using, and storing fissionable materials. Opponents of nuclear power say that the inherent dangers of using fissionable materials outweigh all benefits. But proponents maintain that safety has sufficiently improved, over time, to make nuclear energy an acceptable part of our energy mix.

- Nuclear power plants emit no carbon. For this reason proponents refer to it as clean energy and promote its use as a way to combat global climate change. Opponents of nuclear power do not believe it should be considered "clean" — even though it's carbon-free — because of waste and safety issues.

- With nuclear energy, a huge amount of electricity can be produced from very little fuel. One nuclear power plant can produce hundreds to thousands of megawatts of baseload power.

- In the U.S., nuclear power plant back-up safety systems helped to avoid several nuclear power plant disasters, such as 1979's contained accident at Three Mile Island in Pennsylvania. Nuclear power advocates point to this as demonstrating the effectiveness of the industry. Opponents say that these incidents will continue to occur and show the dangers that might not be avoided the "next time around."

- Additional concerns voiced by nuclear power development opponents is vulnerability of nuclear power plants to sabotage or terrorist attacks.

- Cost factors occasion further controversy. Proponents of nuclear energy argue that its fuel costs are less than for fossil fuel plants, and that domestic nuclear power means less dependence on foreign oil. Opponents respond that costs are high at the beginning (construction) and end (decommissioning); waste management and disposal are costly; substantial water is needed for cooling; and discharge of cooling water can be harmful to marine life.

- The regulatory system in the U.S. includes inspectors at each nuclear plant. It also encourages the power plant operators to identify and correct safety problems. For example, routine examinations of a nuclear facility in Ohio in 2002 found a potentially dangerous situation that caused the Nuclear Regulatory Commission (a U.S. government agency) to order a check of 68 similar nuclear plants. Some say that shows a careful approach. Others say it evidences danger.

- "Spent" rods of used nuclear fuel are typically stored at the power plant that produced them. The U.S. government built a disposal facility and containers for spent nuclear fuel deep under the Nevada desert, but use of this facility has been stalled by controversy. Experts disagree over whether this facility, or others like it that may be built, is an adequate solution for the protection of public safety and the environment.

- A bottom-line question of our time is whether the hazards of nuclear power development and the attendant long-term waste problems are outweighed by the alternatives. A build-up of nuclear power facilities can replace fossil fuel facilities in far less time than it takes to develop other methods of avoiding greenhouse gas emissions. Policy makers must weigh the hazards of nuclear waste into the future against the need to gain control of pollution and climate change now.

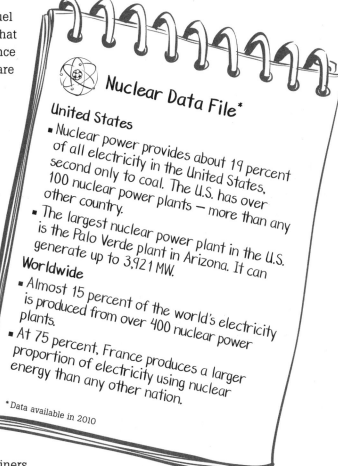

Nuclear Data File*

United States

- Nuclear power provides about 19 percent of all electricity in the United States, second only to coal. The U.S. has over 100 nuclear power plants — more than any other country.
- The largest nuclear power plant in the U.S. is the Palo Verde plant in Arizona. It can generate up to 3,921 MW.

Worldwide

- Almost 15 percent of the world's electricity is produced from over 400 nuclear power plants.
- At 75 percent, France produces a larger proportion of electricity using nuclear energy than any other nation.

* Data available in 2010

ENERGY, HEALTH, AND THE ENVIRONMENT
How energy choices affect our health and the environment

TERMS IN THE GLOSSARY

carbon footprint
carbon monoxide
carbon sink
conservation
ecosystem
encroach
exempt
greenhouse effect
habitat
nitric acid
nitrogen oxides
old-growth forest
organic decay
ozone
particulates
photochemical smog
sediment
sulfur oxides
sulfuric acid
temperate zone
unburned hydrocarbons
wetland

ALL LIVING THINGS NEED CLEAN AIR. They need clean water too, and a temperature range in which they can survive. When the air becomes dirty and polluted, it affects the health of all plants and animals, and it can alter the climate.

Ever since Earth's beginnings, naturally produced pollutants have entered our planet's air from volcanic eruptions, forest fires, dust storms, and pollination. But in the last 200 years, human activities have added greatly to the amount of pollution entering the atmosphere, making it difficult for Earth's natural balancing systems to keep up.

The main cause of excess pollution in our air has been the burning of fossil fuels — for industrial processes, transportation, and electricity generation. Fossil fuel combustion contaminates our air with gases, chemicals, smoke, and ashes — pollutants that are ultimately deposited in our water and soil as well.

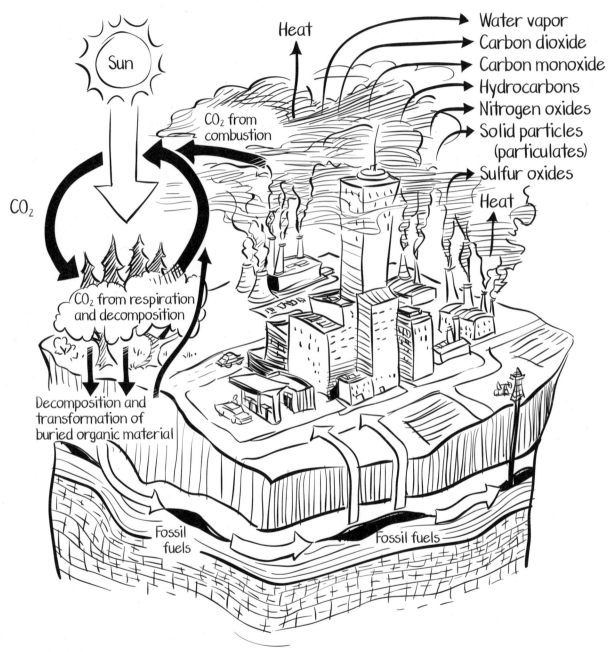

Fossil Fuel Cycle

The fossil-fuel cycle starts with the capture of carbon dioxide by trees, plants, and other vegetation during photosynthesis. Buried organic material goes through chemical changes to form fossil fuels in a process that takes millions of years. The burning of fossil fuels releases heat, water vapor, carbon dioxide, and other air emissions. Some of the carbon dioxide is recaptured by plants, but some can also remain in the atmosphere.

AIR POLLUTION'S HEAVY HITTERS

Pollutant	How Produced	Effects
Carbon dioxide (CO_2)	**In nature:** Forest fires; volcanoes; other natural processes. **By humans:** Burning fossil fuels and biomass.	Excess in atmosphere believed to contribute significantly to global climate change, through the greenhouse effect.
Carbon monoxide (CO)	**In nature:** Forest fires; other natural processes. **By humans:** Incomplete burning of carbon in fossil fuels, reduced by pollution controls.	In upper atmosphere, naturally occurring CO is not a health hazard. At ground level, it is highly toxic, even lethal.
Mercury (Hg)	**In nature:** Volcanoes; oceans; soil erosion. **By humans:** Burning of coal and oil; municipal and medical wastes; mining; cement industry.	Sifts down from the air and accumulates in soil/water; builds up in fish which, when eaten by humans, causes nerve/liver/brain damage; especially toxic to fetuses.
Methane (CH_4) (Natural gas is about 94 percent methane)	**In nature:** Wetlands; peat; termites; oceans; wild animal wastes. **By humans:** Rice farming; natural gas, coal, and biomass production and combustion; landfills; farm animal wastes; human sewage.	Contributes to global climate change and formation of photochemical smog. At higher concentrations, displaces air.
Nitrogen oxides (NO_X)	**In nature:** Lightning; organic decay. **By humans:** Burning fossil fuels, especially coal; certain farming practices.	Contribute to formation of photochemical smog, acid precipitation, global climate change.
Ozone (O_3)	**In nature:** In upper atmosphere, occurs naturally; in lower atmosphere, lightning. **By humans:** In lower atmosphere, formed by a reaction involving sunlight and unburned hydrocarbons produced by burning fossil fuels.	In upper atmosphere is necessary to block the sun's harmful ultraviolet rays. In lower atmosphere is a pollutant causing eye, lung, and throat irritation; also degrades rubber and other materials.
Particulates (Very small particles suspended in the air, including smoke, dust, and vapor)	**In nature:** Forest fires; volcanoes; dust. **By humans:** Burning fossil fuels and wastes; construction; mining; certain farming/ranching practices; winter street sanding.	Can directly harm respiratory tracts, cause haze, damage buildings and other materials; may also contribute to global climate change.
Sulfur oxides (SO_X)	**In nature:** Volcanoes; organic decay. **By humans:** Burning fossil fuels, especially coal, fuel oil, and diesel.	Element of smog that is corrosive and lung-damaging; contributes to acid precipitation that damages lakes, forests, and crops.
Unburned hydrocarbons or volatile organic compounds (VOCs) (excluding methane)	**In nature:** Gas/oil seeps; forest fires; other natural processes. **By humans:** Incomplete burning of fossil fuels; evaporation (fumes) of petroleum fuels, dry cleaning fluids, paints, solvents.	Contribute to formation of photochemical smog.

This chart lists many kinds of air pollution, some caused by nature and some by humans. Much, but not all, of human-caused pollution comes from the burning (or incomplete burning) of fossil fuels.

LOCAL AND REGIONAL AIR POLLUTION

Each of the three main fossil fuels used for electricity generation produces different types of pollutants when burned. Coal puts out the most pollutants: carbon dioxide; particulates such as soot and ash, sulfur and nitrogen gases; unburned hydrocarbons; carbon monoxide; and even small amounts of mercury and radioactive materials. Natural gas mainly produces carbon dioxide and nitrogen oxides. Burning oil results in many of the same gases as coal, but not as many of the particulates.

The American Lung Association estimates the cost of air pollution, in terms of medical care and days lost at work, to be billions of dollars annually. Air pollution affects not only our physical health, but also the economic health of business and industry.

Acid Precipitation

Acid precipitation (commonly, but inaccurately, called acid rain) results when sulfur oxides and nitrogen oxides in the air combine with water vapor to form sulfuric and nitric acids. These acids fall back to the earth in many forms, including rain, snow, fog, or even dry particles. Tall smokestacks on fossil-fuel power plants might seem to help local pollution by dispersing it, but they actually end up putting pollutants higher in the atmosphere, where winds can carry them far away to other communities. Wherever acid precipitation falls, it damages and destroys crops and plant life on land and in water. It can also harm or destroy animal life in these natural habitats.

Photochemical Smog and Regional Haze

There are two main types of smog or haze. *Photochemical smog* forms when sunlight (the "photo" part) reacts with pollutants such as nitrogen oxides and hydrocarbons in the air. This type of smog (an unsightly brown color) can cause lung damage, aggravate asthma and emphysema symptoms, and harm vegetation.

What we call *regional haze* (from sulfate and nitrate particles) scatters and absorbs sunlight, making a clear day smoggy. Regional haze contributes to lung ailments. It also reduces the number of visitors to some national parks, affecting local economies that rely on tourism.

THE GREENHOUSE EFFECT

Greenhouses are warmed buildings used to grow plants. Solar radiation (see page 84) passes through the glass or plastic roof and walls of a greenhouse to the plant-growing space inside. Some of this solar radiation is reflected back from the surfaces inside the greenhouse in wavelengths that are longer, called infrared (heat) waves. These cannot pass easily through glass, so much of the heat stays inside the building.

The earth's atmosphere acts in a similar way. Solar radiation (heat and light) passes through the atmosphere, warming and sustaining life on our globe. Some of the radiation reflects back from the planet's surface.

Greenhouse gases (mainly water vapor, carbon dioxide, methane, and ozone) and particulates in the atmosphere absorb some of the reflected heat and emit some of it back to the planet's surface.

Without the greenhouse effect, our planet would be a cold and lifeless place. However, most scientists believe that excess greenhouse gases in the atmosphere are changing Earth's climate, distorting nature's balancing systems.

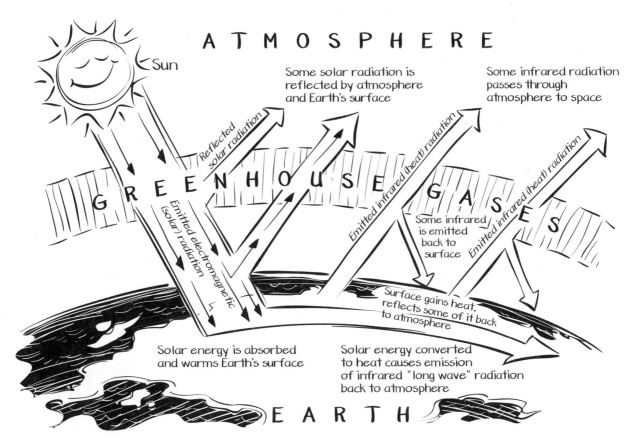

The Greenhouse Effect

SOME AIR POLLUTION SOLUTIONS

Improving the Technology of Pollution Control

We can reduce pollution from electric power plants by requiring their operators to install pollution control equipment and to use fuel as efficiently as possible. Such equipment is available, and scientists are always working on ways to improve pollution control equipment and waste disposal methods. It may be expensive to apply new, cleaner technology to existing, sometimes aging, power plants. To ensure that pollution controls are implemented, local and federal regulations set standards for air quality (although some of the dirtiest, oldest power plants are exempt from these regulations).

Using Clean Energy Sources

Another solution to the air pollution problem is to create less pollution in the first place by tapping clean energy sources. Many of the cleanest are renewable energy sources. Renewable energy sources do not produce the amounts of air pollutants associated with traditional fossil fuel-burning power plants.* Renewable energy sources produce very little or no carbon dioxide when they generate electricity.

Conserving Energy

Another very important way to help avoid air pollution is by conserving (using less) energy. There are many different ways we can do this. (See Chapter 5, pages 143–155.)

GLOBAL CLIMATE CHANGE

Our planet's climate has changed many times. Since the last ice age (about 18,000 years ago) our climate has been gradually warming. In the last 200 years the earth's climate has warmed up at a much faster rate than in earlier years. Some believe that this change is part of a natural series of warming and cooling cycles. Growing numbers think that humans are responsible for the most recent climate changes, and that the long-range effects could be very harmful.

For example, in California alone, the use of renewable energy has kept three million tons of carbon dioxide (that would have come from fossil fuel plants) from being emitted every year. Renewable energy power plants also have prevented the emission of other hazardous air pollutants, including about 17,000 tons of nitrogen oxides and 14,000 tons of sulfur dioxides each year.

YOUR CARBON FOOTPRINT

Nature erases a footprint with ease when it's literally the impression of a foot — but not when "footprint" stands for the effects of modern life. In creating material advances, humans rearrange the environment, especially by moving carbon from Earth to atmosphere when they burn fossil fuels. Since each of us shares that responsibility, we may want to be aware of our individual part in creating green-house gases as our own "carbon footprint."

A carbon footprint is the sum of our primary and secondary footprints. Our primary footprint is a measure of direct emissions of carbon (CO_2) when fossil fuels are burned for the energy needed to heat our buildings, to make electricity for the lights and electrical appliances we use every day, and for the cars, buses, trains, and airplanes we use for transportation. The secondary footprint is a measure of the indirect CO_2 emissions created during the life cycle of products we buy and use, from the manufacturing process to disposal and breakdown. All adds up to emissions on our behalf.

Chapter 5 of *Energy for Keeps* includes suggestions for some of the many ways you can conserve energy and reduce your own carbon footprint.

Calculate your own carbon footprint using one of the calculators at www.carbonfootprint.org.

One thing most people agree upon is that the earth's atmosphere has been warming up, especially since the late 1800s. The earth's average surface temperature set new record highs in recent years. In fact, according to the U.S. National Oceanic and Atmospheric Administration (NOAA), the warmest years on record have occurred since 1990. Temperatures of the uppermost level of the sea have also been rising in many parts of the globe. The Intergovernmental Panel on Climate Change (an international group that assesses scientific climate data) says that this century will experience even greater warming if certain human activities — including how we harness energy sources – are not changed.

The main cause of the current warming trend is thought to be an increase of carbon dioxide and other greenhouse gases (such as methane) in the atmosphere. Experts agree that the burning of fossil fuels for transportation and electricity production is the primary cause. The amount of carbon dioxide in the atmosphere has risen dramatically since the 1850s, when we began to burn fossil fuels in larger and larger quantities. The United States had historically released the greatest share of carbon dioxide to the atmosphere each year — one fourth of all the CO_2 produced in the world — though it has only 4–5 percent of the world's population. In recent years, however, China has been producing more than the U.S.

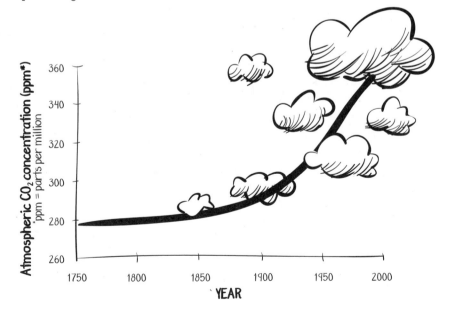

The rise of carbon dioxide levels in the atmosphere

The Potential Effects of Climate Change

In recent years, some natural worldwide patterns and systems have been changing in a dramatic way, enough to attract the attention of international researchers. This notable shift is thought by most climate scientists to be the result of global warming.

For example, along the Antarctic Peninsula, major ice shelves are starting to break up, yielding some of the largest icebergs in recorded history. In the Arctic, sea ice covers less of the ocean than it did 20 years ago, and the ice has thinned. Greenland's ice cap, the second largest ice sheet in the world, is melting, and in Alaska, glaciers are retreating.

As ice sheets and glaciers melt, they add water to the oceans and as ocean waters warm, they also expand. Already, encroaching seas are eroding our coastlines. In some areas, delicate marsh systems are being destroyed by salt from rising seawater. In fact, global sea levels rose an average of 10 times faster in the twentieth century than any time known previously. Even a very small rise can affect millions of miles of shoreline and flood coastal cities. Many people would be affected by this, since much of the world's population lives along the shorelines.

Normal weather patterns will probably be altered by global climate change. Rainfall patterns could change, drought and flooding could be more common, tropical storms could become more severe, and ocean currents and wind patterns could alter. Places that once were warm may begin cooling, and cooler places may warm up. This is why experts prefer the term "global climate change" to "global warming."

Already, habitats worldwide are reacting to changes in climate and weather patterns. Plant and animal ranges are shifting toward cooler latitudes and higher altitudes. Fragile ecosystems, such as those of coral reefs (see sidebar), are also seriously affected. The human habitat may also be impacted. Potential effects are expected to include heatstroke and other health problems, along with economic concerns. These could range from agricultural losses and property damage to decreased tourism. A major economic effect of changing climate conditions is the increased need for energy to respond to hotter- or colder-than-normal temperatures.

RAINFORESTS OF THE SEA

Around the globe, higher sea surface temperatures have been affecting almost all species of corals — sea animals whose hard skeletons eventually form a reef. Coral reefs provide an extensive habitat for varied ocean life, earning them their nickname, "rainforests of the sea." They provide resources for fisheries, medicines, and other products. Many economies rely on them as a draw for tourism. Coral reefs also protect entire island nations from wave and tidal erosion.

Corals need warm waters, but they already live at the upper edge of their temperature tolerance. Therefore, higher-than-normal sea temperatures damage corals, causing bleaching and death. Already, over 25 percent of the world's once-thriving coral reefs are now skeletal grave sites. At current rates, more than two-thirds of the world's coral reefs could be destroyed within the next 30 years, including ancient reefs that have existed for over 1,000 years.

Storing Carbon

Our planet has a natural carbon cycle through which carbon is both produced and stored. Carbon is stored (and thus kept out of the air) during this cycle in carbon "sinks." Two of the primary sinks are plants and the oceans.

Plants use carbon during photosynthesis, removing carbon dioxide from the atmosphere. Old-growth forests are excellent sinks, with their mature trees, other vegetation, and their rich forest floor litter (duff). But all over the globe, old-growth forests are disappearing at an alarming rate. In the United States, most of the old-growth forests are already logged. Tropical rainforests, with their rapid and prolific plant growth, also play a critical role in the carbon cycle. The burning and cutting of these forests worldwide is further disrupting the world's long-standing carbon cycle. In Africa, the Sahara Desert is already expanding onto once-forested areas.

The oceans are the world's largest carbon sinks. Carbon dioxide dissolves in the ocean water. Tiny ocean-dwelling organisms use it to build their shells or skeletons, and so take the carbon with them when they die and fall to the ocean floor. Limestone, a sedimentary rock (primarily calcium carbonate), is formed on ocean floors from the shells and skeletons of these organisms. Some people feel that our vast oceans provide an adequate sink for the excess carbon humans produce. However, there is only so much surface water in direct contact with air, and deep ocean circulation is a slow process. We are increasing the carbon dioxide, but not the oceans' ability to absorb it.

GLOBAL CLIMATE CHANGE SOLUTIONS

Many solutions have been proposed to avoid long-term global climate change (see Chapter 5, "Energy Management Strategies and Energy Policy"), including the following:

- Use energy from renewable energy sources, since they do not produce carbon dioxide. (This includes biomass, because carbon is absorbed while the biomass is growing, which offsets the release of carbon when the biomass is burned.)

- Make the most of the fossil fuels we do use by practicing conservation and energy efficiency. The less we use for a given need, the less carbon dioxide and other greenhouse gases we send into the atmosphere.

- Increase the protection of all remaining old-growth forests and tropical rainforests, while also conserving undisturbed grasslands. Dozens of county and state governments, as well as private conservation organizations, are purchasing large and small tracts of forest, woodlands, and grasslands so they will be preserved in their natural state.

- Practice "restoration ecology," which includes revegetating areas with the plants and trees that originally grew in a certain locale before it was disturbed by human activity.

- Research other ways (besides natural processes) to capture more carbon dioxide in the future. Some experimental ideas include injecting it into underground oil or coal fields or using chemical methods to bind it with other substances.

- Enforce air pollution standards. Policies that reduce overall air pollution levels also cut carbon dioxide emissions and are beginning to become more widespread. Even some private industries, including several fossil fuel industries, have begun to reduce their carbon dioxide emissions and invest in renewable energy sources. Insurance companies, fearing the huge claims that might arise from climate change, are also speaking up.

- Support international cooperation. Over 100 countries have signed greenhouse gas reduction agreements, even though there still exists an ongoing debate about the causes of global climate change and its effects. Recent revisions in government policy and in corporate attitudes may indicate that environmental issues such as these are being taken more seriously.

ENERGY MANAGEMENT STRATEGIES AND ENERGY POLICY
How energy decisions affect our lives

TERMS IN GLOSSARY

active solar

American Recovery and
  Reinvestment Act (ARRA)

blackout

brownout

Clean Air Act

compact fluorescent light bulb (CFL)

cool roofs

demand response

direct use geothermal

distributed generation

ecological

energy efficiency

green pricing

Green Tags

incandescent light bulb

independent power producer

indirect (hidden) costs

microgrid

net metering

passive solar

peak load

policy

rebate

renewable energy certificate (REC)

renewable portfolio standards (RPS)

smart grid

tradeable renewable energy credit
  (TRC, TREC)

U.S. Environmental Protection
  Agency (EPA)

true-cost pricing

REMINDER

W = watt
kW = kilowatt = 1,000 watts
MW = megawatt = 1,000 kilowatts
1 megawatt can serve about
1,000 homes in the United States.

WHEN IT COMES TO ENERGY, you can make a difference, whether you are a student or the president of a country. Each individual counts when it comes to energy use. If five million of us turned off just one unneeded light all at the same time, for instance, we would reduce the demand for electricity by about 500 MW. This is the size of a typical large power plant. And during the summer, if only one family or small business adjusted its air conditioning thermostat up by 3°F, or about 2°C, it would keep about 470 pounds (213 kilograms) of carbon dioxide from being emitted into the air every year.

In many ways we are all connected to each other and to the environment in which we live. Our energy decisions affect our own quality of life and the lives of others. As individuals and collectively, we can pursue energy choices that benefit everyone.

CONSERVING ENERGY AND INCREASING EFFICIENCY: EVERYONE CAN MAKE A DIFFERENCE

We often hear about the need to save, or conserve, energy. This doesn't mean not *using* energy. It means not *wasting* energy. There are several important reasons not to waste energy. One is that, since much of our electricity is produced using nonrenewable fossil fuel resources, we want to make the best use of the fossil fuels we have left and save some for future generations. Another reason is that the less fossil fuel we use, the less pollution we produce.

Your home is a good place to start conserving energy. The biggest uses of energy in an average American home are for heating and cooling. These represent about 44 percent of energy used. Refrigerators consume almost 10 percent, and lighting, cooking, and other appliances use approximately 33 percent. Water heaters use up most of the rest. Buying appliances that are energy efficient is one way to reduce energy waste. (Perhaps it is time to replace that old energy-gobbling refrigerator.) However, there are many other low- or no-cost energy savers. These include turning up your thermostat several degrees in summer (if you have air conditioning) and down a few notches in winter; changing furnace air filters frequently; using shades or curtains to block the sun in summer; using compact fluorescent light bulbs; turning off lights, TVs, stereos, and computers when you aren't using them; and installing insulation in your attic and walls.

PEOPLE POWER

The people of Princeton, Massachusetts, have demonstrated more than once that ordinary citizens can make a difference in the quality of life in their community. In the mid-1980s, in order to meet an increasing demand for electricity, their local utility proposed buying electricity from a nearby nuclear power plant. Princeton residents voted instead to build their own wind farm, which they supported with a bond measure. Eight wind turbines were installed on a windy ridge of Mount Wachusett in 1984. Then, in 2003, Princeton residents once again spoke in favor of renewable energy. In another election, they voted to replace the eight older, less-efficient wind farm turbines with two tall, sleek, 1.5-megawatt turbines. In spite of some concerns about the "look" of the taller turbines, the majority of Princeton voters were proud to produce their own clean electricity from a local renewable resource.

Another good energy-saving solution is to plant deciduous trees — trees that lose their leaves in winter. If you plant them along the sunny sides of your house (or ask your building manager or owner to consider doing so), you will have shade on those hot sides of your house in summer. In the winter, with the leaves gone, sun can reach through the bare branches to warm your house.

An Energy Efficient Home

Everyone can save energy at home; many methods and materials are shown below. Other approaches to conserving energy include the use of solar water heating systems and geothermal heat pumps. (See page 149.)

COOL ROOFS

Air conditioning makes a hot day the time of highest energy use. That's when peaking power, the most expensive electricity, is used — so that's when efficiency in building construction is most cost effective. Heat comes through the roof, raising the temperature of a dark roof to levels that can exceed 150°F (66°C). A light-colored roof will reflect heat and reduce the surface temperature dramatically, sometimes by over half. A cooler roof means cooler interiors and less electricity use.

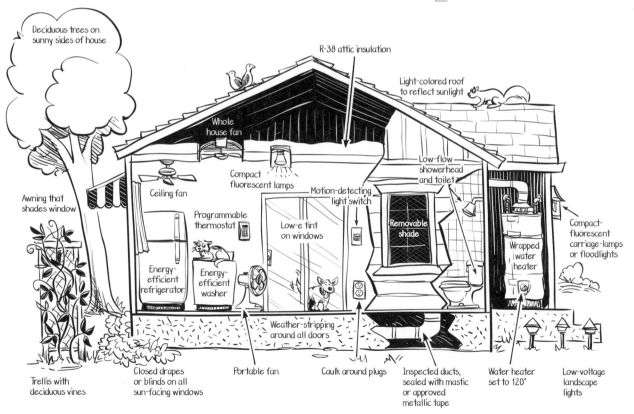

Deciduous trees on sunny sides of house

R-38 attic insulation

Light-colored roof to reflect sunlight

Whole house fan

Compact fluorescent lamps

Motion-detecting light switch

Low-flow showerhead and toilet

Ceiling fan

Awning that shades window

Programmable thermostat

Low-e tint on windows

Removable shade

Compact-fluorescent carriage-lamps or floodlights

Wrapped water heater

Energy-efficient refrigerator

Energy-efficient washer

Weather-stripping around all doors

Trellis with deciduous vines

Closed drapes or blinds on all sun-facing windows

Portable fan

Caulk around plugs

Inspected ducts, sealed with mastic or approved metallic tape

Water heater set to 120°

Low-voltage landscape lights

Permission given from Sunset Publishing Corporation to adapt graphic from *Smart Water and Energy Use in the West*

WHAT'S THE BIG DEAL WITH ENERGY EFFICIENCY?

❶ Energy spent to mine coal

❷ Energy spent to transport coal

❸ Energy spent to process coal

❹ Chemical energy (coal) goes in

❺ Thermal energy spent to make steam

❻ Mechanical energy spent to turn turbine-generator

❼ Heat energy lost in condenser

❽ Heat energy lost in transmission lines.

❾ Heat energy lost from use of light bulb

Pump   Condenser   Cooling Tower

Energy efficiency is never 100 percent.

When a device performs work for us, some of the energy is lost (wasted) as heat. Some processes and appliances waste more energy than others. In a standard, incandescent electric light bulb (the kind some of us still use at home) only 5 percent of the energy entering the bulb is converted to light! Most of the energy is lost as waste heat from the filament that glows to produce the light. (Compact fluorescent light bulbs are up to 75 percent more efficient and last 10 times longer.) Older models of refrigerators, other appliances, and gasoline and electric motors also waste a great deal of energy. Newer models use less, but still get the job done.

An energy system — such as a power plant — can also be thought of in terms of efficiency. At each step in such a system — from creation to use — some energy is lost, mostly as heat. Say, for example, we light that incandescent light bulb with electricity from a traditional coal–fired power plant. During the many steps required to make electricity at such a plant (mining, transporting, processing, burning, spinning the turbine and waste disposal), about 65 percent of the coal's energy is lost. Next, about 10 percent more of the remaining 35 percent is lost as heat when the electricity moves along transmission lines. Then the electricity is used to light our inefficient light bulb. That's a lot of wasted energy! Luckily, many of our renewable energy technologies use more energy efficient systems.

IMPROVING GRID EFFICIENCY

Electric power grids have been in place as long as telephone systems have, but advances for electricity have come more slowly. Our aging grids don't respond efficiently to part-time electricity sources such as sun and wind. Upgrading to "smart" grids and "micro" grids can save energy, reduce cost, and increase reliability.

Smart Grids

New "intelligent" two-way tracking systems can share information about electricity flow among power providers, grid operators, and consumers. That will conserve grid space and improve use of generation capacity. A *smart grid* will be able, in the future, to send power to and from advanced storage devices, which can include electric fuel cell cars.

Microgrids

Large size can be a disadvantage for a grid. Distance brings power loss. And disruptions, whether by bottleneck or by malfunction, can cause widespread effects, even brownouts or blackouts. An alternative is to use *microgrids*, small and operated locally. A microgrid can be linked with other grids for shared power; or it can be "islanded" (walled off) for protection from others' disruptions. Another important use is in remote areas with no nearby large grid.

DISTRIBUTED GENERATION

Another way to make energy more efficient is to generate power closer to where it will be used. Distributed Generation (DG) avoids the long distance transmission line power losses — which can be as high as 15 percent. The first DG photovoltaic and wind systems were installed in remote off-grid locations. DG is also now supplied to the grid by small power plants built to serve particular large businesses, or sited near the end of distribution lines. It is also useful as back-up power for hospitals, television stations, internet servers or other critical facilities. These mini power plants can be energy-efficient natural gas micro-turbines, small hydropower plants, modular binary geothermal units, photovoltaics, landfill gas, wind turbines, or banks of fuel cells. They are especially well suited for microgrids.

An example of distributed generation: microbe-eating bacteria in the adjacent landfill produce methane as fuel for these micro-turbines.

THE RIGHT SOURCE FOR THE RIGHT USE

One way to get the most out of the energy we use is to match the
right source with the right use. For example, if we need heat, then
we can use a source that is already hot, such as solar or geothermal
energy. Or, we can use the waste heat from power plants or industrial
processes. That way we don't need to use electricity or burn fuel to
produce heat. These approaches save resources, are energy- and cost-
efficient, and are easier on the environment. Here are some of the
ways we do this.

Active and Passive Solar Heating

Active solar heating systems absorb heat using solar collectors (often
found on rooftops) that are filled with water or another liquid. Pumps
can move the heated liquid through equipment to warm a building,
take the chill from a swimming pool, or preheat water for a water
heater. Buildings with *passive* solar systems naturally collect the sun's
heat during the day, using thick walls or large tile or brick floors in
the sunny areas, and expansive south-facing windows. Rooms with
these features are sometimes called sunspaces. These "solar collectors"
then slowly release the heat at night, when temperatures dip and the
heat is usually needed. In summer, deciduous trees or awnings can
shade the windows or walls from the sun.

PORTABLE POWER IN AFRICA

I n Africa's Sahara Desert, health care workers needed
a way to carry perishable vaccines across the
hot, sandy desert to remote health clinics.
The workers decided to make use of the
best options available to them: camel
caravans and the sun. The vaccines
traveled in style, safely carried in small
refrigerators powered by solar charged
batteries — all sitting atop the camels!

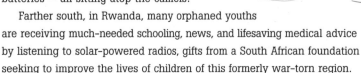

Farther south, in Rwanda, many orphaned youths
are receiving much-needed schooling, news, and lifesaving medical advice
by listening to solar-powered radios, gifts from a South African foundation
seeking to improve the lives of children of this formerly war-torn region.

Direct Use Geothermal

Geothermal water that is not hot enough for electrical generation can still be very handy. Around the world it's used "directly" in many different ways. After pumping the water up from geothermal reservoirs, people use it to heat homes, schools, offices, greenhouses, swimming pools, and even water for fish farms. In some places, hot geothermal water is run through pipes under sidewalks and roads to melt snow and ice in winter. It also provides heat for industry, where it's used to dry agricultural products and lumber.

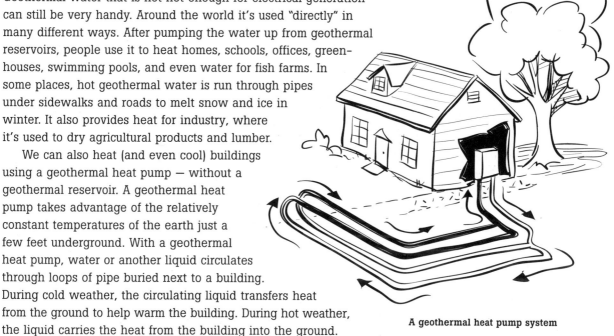

A geothermal heat pump system

We can also heat (and even cool) buildings using a geothermal heat pump — without a geothermal reservoir. A geothermal heat pump takes advantage of the relatively constant temperatures of the earth just a few feet underground. With a geothermal heat pump, water or another liquid circulates through loops of pipe buried next to a building. During cold weather, the circulating liquid transfers heat from the ground to help warm the building. During hot weather, the liquid carries the heat from the building into the ground.

Cogeneration

Cogeneration, also referred to as combined heat and power (CHP), is another way to use energy more efficiently — reducing costs, saving energy, and cutting back on pollution. It means producing electricity and heat at the same time, from the same fuel or energy source.

Facilities using cogeneration produce their own electricity and then use the resulting "waste" heat for another use. For example, a pulp and paper mill might produce its own electricity using a steam-driven turbine-generator, then use the waste heat to help produce paper products.

If it's a power plant that uses cogeneration, the waste heat that is captured after producing electricity can be used to produce even more electricity, to heat power plant offices, or to be sent right next door for use at a fruit-drying plant, for example.

A wide variety of power plant types can make use of cogeneration, including geothermal, solar thermal, landfill gas and other biomass plants, hydrogen combustion, fossil fuel, nuclear, or even fuel cell power plants. A "combined cycle" power plant (see page 122) is actually a cogeneration power plant.

ENERGY POLICIES: DECIDING WHAT'S IMPORTANT TO US

Policies are the guidelines or principles that we use when we make choices. They can be simple or complex.

Our personal policies reflect our own individual beliefs and goals; they are relatively simple. Policies for groups are far more complicated. They must reflect shared goals arrived at from differing interests and points of view. Whether for a family, a business, a school, a community, or a nation, policy-making requires shared information, consideration of needs, and compromise.

Individuals, businesses, and governments can all have their own energy policies, but it is the energy policies of state and national governments that affect our lives the most. Governments have a big say in which energy sources are developed, how much energy is imported, and even how much we pay for energy. Government energy policies affect public health, the environment, the economy of a region, the security of a nation, and the energy choices available to all of us and to future generations.

Government Power Policies

Before the mid-1970s, most of the electricity in the U.S. came from large centralized power plants owned by huge utilities regulated directly by government agencies. These utilities generated all the electricity within their service territories. They were the only ones allowed to build power plants. There was no competition. When our imported oil supplies were threatened in the 1970s, Congress wanted to encourage the development of new energy sources and greater energy efficiency, and businesses that consumed a lot of energy also began demanding more choice in power providers.

Federal laws were passed in 1978 requiring certain utilities to use electricity from independent producers. The electricity market was opened up for these independent power companies, many of which began producing electricity from renewable energy resources. Though regulations have evolved, independent power producers remain and have become a permanent part of our energy scene.

More than half of new electric generation in the U.S. now comes from these independent power producers. This allows many utilities to offer their customers choices that include electricity from renewable energy sources. Sometimes the utilities buy this electricity from independent power producers, and sometimes they generate their own.

DEREGULATION

Deregulation literally means elimination of regulation from a previously regulated industry. Many states are deregulating the electricity and natural gas industries to make electricity sales less a utility monopoly and more an open market. But it doesn't mean there are no rules. It just means the rules are changing. One common way to deregulate, also known as restructuring, is to make it possible for large businesses, a group of small customers, or an entire city to join together to buy their electricity directly from a supplier rather than from a regional utility. (See "A Local Voice on Electricity Choices," page 154.) The electricity industry in the U.S. will continue to change as new models are tried and either work or fail.

WEIGHING IN ON FEDERAL POLICY

A balancing act of decision-making goes on every working day for our elected officials in Congress. Public health and the environment must be balanced with economic concerns and other interests of voters. You can influence these public policy decisions by contacting your representatives in Congress to let them know what's important to you.

One important example to consider is the Clean Air Act (CAA). Originally passed in 1970, the CAA is a Federal law designed to protect public health. It requires the U.S. Environmental Protection Agency (EPA) to set air pollution standards and enforce regulations for industry and power plants.

Some industries and energy producers have not liked this regulation. They say it costs too much to use all the pollution controls required. However, other businesses have found that by using energy efficient measures and cleaner energy sources, they have not only met pollution standards, but have actually increased their profits.

Some people have pressed to repeal the CAA. Others wish to keep it in place, but would like to change the specific ways the regulations are put into action. Over the years, however, the basic intentions of the CCA have remained in place.

Policies such as these have had notable success in controlling pollution. For example, air quality in the Los Angeles area of southern California has improved dramatically due to the efforts of local agencies, such as the Air Quality Management District, which makes sure that both the CAA and state pollution regulations are being enforced.

With a complex issue such as the CAA, members of Congress need to be informed about all sides. The opinions of their citizenry — young and old — add weight to one side or the other of an argument. Therefore it is in everyone's own best interest to be informed and vocal about what happens at our nation's capital, especially when such important matters hang in the balance.

IMPLEMENTING ENERGY MANAGEMENT STRATEGIES

Energy policies are implemented by an energy management program — plans of action, or strategies. In this section you will find examples of energy management strategies, some of which can be pursued or are already being implemented by you, your family, business, school, utility, or local, state and federal governments.

Shifting the Load

There are times when everyone in a region uses a lot of electricity, all at once. You'll recall from Chapter 2 that this is called a peak load. In general, people use more electricity (especially in the summer on particularly hot days) between 12 noon and 6 P.M.— the peak hours. Sometimes there just isn't enough electricity for everyone, all at the same time. For example, this can happen during a heat wave when everyone wants to run air conditioners. Because of stresses on the electricity grid, certain areas may end up going without electricity for a specific period (usually several hours), which is commonly called a blackout. Blackouts or brownouts (meaning reduced power) may also occur when something damages the production equipment or wires.

We can all pitch in to help prevent blackouts and brownouts by using our major appliances at "off-peak" times when possible. We can also pay closer attention to how much we use our air conditioning and heaters.

Time of Use Pricing

To encourage load shifting (i.e., shifting the times customers use their electricity), some power providers can now install "time of use" meters that record both the kilowatt-hours used and the time of use. Customers are then charged more for electricity used during peak hours (when the cost of electricity is higher to the power provider and the most-polluting plants are turned on). Customers with time of use metering or rates can reduce their bills by doing laundry, for instance, during off-peak hours. Some utilities go further to spread their power efficiently. A utility in Idaho, for example, pays ranchers to turn off their electric water pumps at peak hours.

Net Metering

This is a program that encourages residences and business to generate some of their own electricity from specified renewable sources like solar PV or wind. If consumers need more electricity than they can generate at any point in time, they can draw on energy from the grid. If they generate an excess of electricity, some utilities buy the excess.

Incentives for Renewables

A number of state energy agencies and utilities give cash rebates for the purchase and installation of certain renewable energy systems, such as PV, small wind, fuel cells, and small biomass systems. Some cities even give homeowners the money to pay up front for such projects and let them pay it back through installments on their property tax bills.

ENERGY STAR Program

The U.S. Environmental Protection Agency established ENERGY STAR in 1992 to identify and promote energy-efficient products. ENERGY STAR-rated products have superior energy efficiency and now include office equipment, home appliances, heating and cooling equipment, lighting, home electronics, and even new homes and offices.

Efficiency Standards and Assistance

Federal and state governments also establish energy efficiency standards — laws setting standards for new products or for entire buildings. There are also programs that assist people who want to make older homes or other buildings more energy efficient. Even government buildings are being made more efficient under these programs.

Demand Response

Transmission line operators and utilities must sometimes respond quickly to unanticipated changes in electricity demand often caused by the weather, breakdowns at power plants, or stresses on transmission line capacity. "Demand Response" is a method for responding to such changes. It simply means that the utility calls customers (usually large businesses for factories) and asks them to reduce their electricity use. The technology is now available to implement Automated Demand Response technologies (AutoDR), which would make it possible to reduce a customer's electricity use automatically (with their prior permission), with control in the hands of the utilities.

TRUE-COST PRICING

Traditionally, the price of a product reflected only the cost of making and delivering it. If a factory polluted the air and people got sick, those costs and losses were borne by the community, not the company that did the damage. These are called hidden, or "external" costs.

Electricity also has hidden costs. Health and environmental costs, for example, result from burning fossil fuels. Air pollution causes ailments, especially of the lungs, which can require costly treatment and lost work days. And there are risks in importing fuels, the costs of which are covered in military and defense budgets (our tax dollars), not in our electric bills.

With renewable energy, we get added value without paying for it — such as reducing the cost of waste disposal to landfills by using biomass for electricity, improving our "balance of trade" by importing less fuel, and increasing our energy security by using more numerous, small distributed power plants.

When the true costs are considered, renewable energy can be more cost-effective than electricity from fossil fuels. Understanding this, consumers are often willing to pay a bit more for renewable energy, knowing they are actually getting good value.

Green Tags and Green-E

A widely used program to promote cleaner energy uses renewable energy certificates (RECs), also known as tradeable renewable energy credits (TRECs) or "green tags." When a utility or customer buys RECs, the purchase price goes to the construction of renewable power facilities, either locally or at a distance. The task of determining if a plant is truly renewable and qualified for REC treatment is delegated to a number of agencies — such as California's Center for Resource Solutions, which administers the Green-E program.

Renewable Portfolio Standards (RPS)

Most states are adopting standards for the minimum amount of renewable energy that must be included in the utilities' mix (portfolio) of electricity resources. As of 2010, more than half of the states have enacted laws to specify that a portion of their electricity must come from renewable energy. (States can define renewables differently.) In some states, the standard can be met by a utility purchase of green tags. Many environmentalists and business leaders are also working to have a set of national renewable portfolio standards established in the near future.

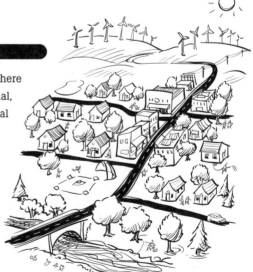

A LOCAL VOICE ON ELECTRICITY CHOICES

Cities and counties in a few states can decide for themselves where to get their electricity. They can buy electricity from geothermal, wind, solar — or whatever they want — without forming a municipal utility. Under this "community choice aggregation" (CCA), a local agency buys power from the producer it chooses, and the area's utility is required to deliver it for a regulated charge.

If the agency chooses a less polluting sources than the utility was providing, the CCA process can allow a community to help clean the air. In Ohio, for example, 200 small towns "aggregated" their power purchases and cut harmful emissions by an estimated 50 percent.

Feed-In Tariffs

Many countries in Europe and several U.S. states are implementing a "feed-in tariff" policy. Under a feed-in tariff program, a utility must buy all renewable power offered (if it meets established rules) at a price set by a governing agency. In some cases, even if the utility has to shut down fossil fuel plants in order to take the renewable power, they must do so or pay a penalty. This policy provides a guaranteed market for renewable power developers. It helps keep their costs down by avoiding expensive bidding competition and it gets more renewable electricity into the grid.

Government Research and Tax Policies

Government assistance has had a large influence on growth in U.S. energy production. Sometimes the assistance takes the form of research and development funds; examples are U.S. Department of Energy (DOE) programs to promote cleaner coal, safer nuclear technology, and more efficient renewable production. Other assistance has been in the form of tax benefits; an example is the Production Tax Credit, which has assisted renewable energy developers.

In 2009, assistance to energy development was taken to a new level by the American Recovery and Reinvestment Act (ARRA). ARRA provides a temporary 30 percent tax credit for developers of specified renewable energy technologies. Further, for the benefit of developers who don't owe that much tax, ARRA provides that (on completion of a qualifying project) the U.S. Treasury will pay cash in the amount of the credit.

Tax benefits for specific industries are controversial. Lawmakers disagree over what projects deserve benefit, and some promote a strictly free market approach in which the government avoids picking winners. The result of the controversy is a package of tax compromises, enacted into legislation once every two years or so, that benefit favored energy sources, usually for a limited time.

ECOLOGICAL FOOTPRINTS

When we walk on a beach or in the snow we leave footprints. Less visible, but much more important, are the ecological footprints we leave when our activities alter the environment or result in the overuse of our natural resources. An ecological footprint is a measure of how much of nature's resources we use to sustain our lifestyles. We all have footprints; however, some are far bigger than others.

Our footprints grow as the economy, the world's population, and our use of natural resources grow. Sometimes the resources we use are renewable — like the trees that supply the wood for building houses or for biomass energy. In other cases — for instance, the consumption of oil — the resources diminish over time. Either way, our footprints may become permanent if we exceed nature's ability to regenerate itself.

We have choices to make —
about how much electricity
we use, when we use it, and
the energy resources we use
to generate it. These choices
are becoming more important
every day.

—Chapter 2, *Energy for Keeps*

Appendix and Index

Energy Timeline 159

Glossary 165

Additional Information Resources 177

Index 185

The Authors 191

ENERGY TIMELINE

4 million B.C.
First known use of tools in East Africa (muscle power)

460,000 B.C.
First known use of fire in area now known as China

10,000 B.C.
Asphaltum from natural oil seeps used for variety of purposes on America's Pacific coast

9000 B.C.
Farming begins in the Middle East and elsewhere; people begin permanent villages

6500 B.C.
Metalworking with copper begins in Middle East

3500 B.C.
Sailboats powered by wind used on the Nile in Egypt

3200 B.C.
Wheels used in Urak, Mesopotamia

3000 B.C.
First recorded use of crude oil, in Mesopotamia

2000 B.C.
Chinese use crude oil for home heating

1500 B.C.
Hittites (Asia Minor) first produce wrought iron

1500 B.C.
Fire-starting kits carried in Europe

1500 B.C.
People around the world use hot springs for bathing, healing, recreation, cooking, heating

1000 B.C.
Iron becomes commonly used metal throughout Mediterranean

750 B.C.
Ironworking reaches Europe

500 B.C.
Magnetic properties of lodestone (type of iron) described by Thales of Miletus in Greece

500 B.C.
Iron plow share first used in Europe, improving the efficiency of plowing using muscle power

500 B.C.
Passive solar energy used in Greek homes

200 B.C.
Coal mining in China

50 A.D.
Hero of Alexandria invents first steam engine (but does not put it to productive use)

50
Romans make wide use of solar energy; improve glass windows

100
Greeks invent waterwheel

300
Natural gas drilling in China

644
First windmill with a vertical axis, recorded in Iran

700
Iron smelting introduced in Spain

1060
Possibly world's first city-wide space-heating project using geothermal built at Paquimé, Mexico

1088
Water-powered mechanical clock made by Han Kung-Lien of China

1100
Oil wells drilled in Europe and the Mediterranean

1100
Windmills introduced in Europe

1200
Coal mining begins in England

1320
Germans improve blast furnace, advancing the process of iron smelting and casting

1322
French village pipes water from hot springs for home heating

1400
Blast furnace introduced in Holland, enabling the first production of cast iron in Europe

1510
Leonardo da Vinci designs the precursor of the water-driven turbine

1582
Waterwheels first pump water from Thames River into mains in London

1615
Use of coal for heating in England increases, owing to rising timber costs

1680
Mills driven by waterpower in common use throughout Europe

1688
Large windows allow solar energy into buildings in France

1690
Widespread use of coal begins in Europe due to wood depletion

1695
Frenchman G. Buffon uses mirrors to concentrate sunlight to burn wood and melt lead

1698
Englishman T. Savery develops steam engine to pump water out of flooded coal mines

1700
Textile mills and other factories driven by waterpower throughout Europe

1700
Greenhouses with glass windows that take advantage of solar energy become popular

ENERGY TIMELINE (continued)

1705
T. Newcomen, England, invents first practical steam engine

1709
Iron smelting process using coke developed by A. Darby, England; coal demand increases

1712
Piston-operated steam engine built by T. Newcomen

1746
B. Franklin conducts research that will later result in clearer understanding of electricity

1748
First American commercial coal production in Virginia

1752
B. Franklin's kite experiment verifies nature of static electricity; leads to invention of lightning rod

1757
First public gas streetlights in the American colonies light streets of Philadelphia

1769
Improved steam engine patented by J. Watt, England

1770
Spinning jenny patented by J. Hargreaves helps automate manufacturing

1782
J. Watt invents rotary steam engine; soon to have widespread use in factories

1785
Textile plant in England is the first to be powered by steam

1790
First working United States cotton mill

1792
British engineer W. Murdock invents "town gas"

1800
A. Volta produces the first electricity from a wet-cell battery

1800
Several French towns use geothermal energy for space heating

1800
Hot springs resorts flourish throughout United States, Europe, and Asia

1803
Robert Fulton builds first steam-powered boat

1804
R. Trevithick invents and operates first steam locomotive on a track

1807
Commercial paddle-wheel steamship cargo service begins in New York

1807
First public street lighting using town gas occurs in London

1814
First practical steam locomotive invented by G. Stephenson

1818
First steamship (*Savannah*) crosses the Atlantic

1820
A. Ampere, M. Faraday, and W. Sturgeon experiment with electromagnetism

1821
M. Faraday, England, demonstrates that electricity can produce motion

1821
First U.S. natural gas well drilled in Fredonia, New York

1825
First steam train passenger service offered in England

1830
Steam-driven cars common in London

1831
J. Henry perfects electric motor

1831
M. Faraday invents dynamo, one of the first electric generators

1839
Englishman W. Grove builds first fuel cell

1859
First oil well in America drilled in Pennsylvania

1860
First internal combustion engine built by E. Lenoir, Belgium

1860
The Geysers, California, opens resort for therapeutic hot spring bathing

1861
French scientist A. Mouchot patents world's first solar steam engine

1868
First modern focusing solar power plant heats water for steam engine in Algiers

1870
Z. Gramme perfects dynamo, making it the first workable electrical generator

1874
Power plant in England burns garbage (biomass) for electricity production

1876
N. Otto perfects first practical internal-combustion engine, later used in autos

1876
California's first "commercial" oil well drilled near Newhall, California

1878
T. Edison develops method to transmit electricity for common use

ENERGY TIMELINE (continued)

1879
T. Edison makes incandescent electric light practical

1881
J. d'Arsonval originates idea of using ocean as energy source

1882
Electric power stations go on-line in London and New York

1884
C. Parson develops first practical steam turbine electricity generator

1885
C. Benz develops the first working motorcar powered by gasoline

1886
Swede J. Ericsson invents first parabolic trough solar energy collector

1886
Up to 50 small hydropower plants generate electricity in America

1887
Stockton becomes first California city supplied with natural gas sent through pipelines

1888
First wind machine for electricity built in America

1890
Electricity begins to replace use of natural gas for lighting

1890
First dependable electric motor cars developed in France and Great Britain

1891
U.S. inventor C. Kemp patents first commerical solar water heater

1891
Tesla coil invented, producing first high-voltage electricity

1891
First long distance electrical lines completed in Germany

1892
P. LaCour, Denmark, designs efficient machine that generates electricity from wind

1893
First Ford gasoline buggy driven by inventor, H. Ford

1894
Texas oil discovered while drilling for water

1894
Pneumatic (air-filled) tires introduced in France by A. and E. Michelin

1896
First U.S. offshore oil wells (built on wooden piers) drilled near Summerland, California

1896
Niagara Falls hydropower plant sends first long distance electricity in U.S.

1897
C. Parsons outruns every ship in the water with his steam-driven boat, *Turbinia*

1897
30 percent of homes in Pasadena, California, use solar water heaters

1898
Garbage (biomass) burned for electricity in New York

1900
Power plants driven by hydropower or fossil fuels dot the U.S.

1900
Calistoga, California, hosts over 30 hot springs resorts

1904
Electricity generated from geothermal steam in Larderello, Italy

1905
A. Einstein publishes relativity theory, revolutionizing understanding of energy

1908
First cheap, mass-produced car, the Model T Ford, is available

1910
Coal accounts for three-fourths of all fuel used in United States

1916
Einstein's unifying theory inter-relates mass, energy, magnetism, electricity, and light

1918
Denmark produces electricity from over 100 wind generators

1920
Midwest farms in U.S. widely use wind turbines for electricity

1920
Decade begins with oil and gas shortages in California

1928
More than 3 million American families own two cars

1929
After major discoveries, decade ends with surplus of oil and gas in California

1930
Iceland begins to work on large-scale geothermal district heating project

1930
Solar water heaters supply hot water to homes throughout Miami, Florida

1930
Propeller-type wind generators perfected by M. Jacobs in use all around U.S.

1932
Englishman F. Bacon develops first successful fuel cell

ENERGY TIMELINE (continued)

1935
Rural electrification brings power to remote areas in U.S.; replaces most wind turbines

1936
America's Hoover Dam (for hydropower) completed

1939
Europeans O. Hahn, and L. Meitner unveil process of nuclear fission for energy

1940
First U.S. superhighway opens in Pennsylvania

1941
Almost 60,000 solar water heaters in use in Florida

1942
E. Fermi, using Einstein's theories, produces first controlled nuclear chain reaction in the U.S.

1943
132 MW produced from geothermal fields, Larderello, Italy

1944
U.S. National System of Interstate Highways established

1945
First nuclear bomb detonated in New Mexico

1945
5,000 U.S. homes have television sets

1947
Diesel-electric trains replace steam locomotives in U.S.

1948
One million U.S. homes have television sets

1950
Work-saving appliances and tools use increasing amounts of energy

1952
First U.S. hydrogen bomb detonated with 700 times force of fission bomb

1954
First solar cells used for electricity generation developed in U.S.

1954
First Russian nuclear power plant opens

1954
Advanced European steel-manufacturing method introduced in Detroit

1954
First fuel cells used in NASA space program

1955
First U.S. town powered by nuclear energy (pilot project) in Idaho

1958
First major offshore oil-drilling platform built in the Pacific Ocean near Summerland, California

1960
Commercial electricity first produced from geothermal energy at The Geysers, in California

1960
Environmental concerns increasingly relate to energy use and pollution

1960
German U. Huttrer perfects electrical wind turbine design, later adopted in U.S.

1963
First commercial nuclear power plant in U.S. opens in New Jersey

1965
Historic electrical blackout in northeastern North America

1966
Partial meltdown at nuclear power plant in Detroit

1966
La Rance tidal power plant built at the Rance estuary in France

1967
First microwave for home use introduced

1968
78 million U.S. homes have television sets

1969
France begins large district-heating projects with geothermal energy

1970
First Earth Day signals worldwide concern about environmental damage

1970
Solar water heating well established in Israel, Japan, Australia

1971
P. McCabe, Great Britain, and M. McCormick, U.S., begin development of first wave energy system

1973
Oil embargo opens up new era of electricity produced from renewable sources in U.S.

1973
Japan begins experiments with Ocean Thermal Energy Conversion (OTEC)

1974
J. Lindmayer, U.S., develops silicon photovoltaic cell for harnessing solar power

1977
Solar panels installed on the White House

ENERGY TIMELINE (continued)

1978
Public Utility Regulatory Policies Act, PURPA, requires utilities to buy power from qualifying independent producers

1979
Partial meltdown of nuclear reactor at Three Mile Island, Pennsylvania

1979
Experimental OTEC project begins producing electricity in Hawaii

1980
Europe and Asia invest widely in generation of electricity from wind power

1980
Nuclear power generates more electricity than oil in U.S.

1980
Large, powerful wind generators emerge as result of fuel shortages

1982
Solar One power tower in southern California proves that solar thermal power for electricity is feasible

1983
Three out of every four power plants in U.S. burn fossil fuels

1983
World's largest hydroelectric power plant opens in Brazil/Paraguay

1983
First solar thermal trough power plant opens in southern California

1984
Large scale biomass power plant opens in Vermont

1986
Worst-ever nuclear meltdown with nuclear fallout occurs at Chernobyl, Ukraine

1990
More than half of world's wind-generated electricity produced in California

1993
Nuclear power provides about one-fifth of U.S. electricity

1997
Hydropower now produces only 10 percent of U.S. electricity

1999
U.S. consumption of petroleum reaches all-time high, more than half for transportation

2000
Electricity generation produces almost 40 percent of all carbon dioxide emissions in U.S.

2000
Of the carbon dioxide emissions produced from electricity generation in the U.S., over 80 percent are from coal

2000
Injection of wastewater into The Geysers geothermal reservoir boosts electricity production

2000
Utility deregulation in some U.S. states results in ups and downs in opening up the energy production market

2000
State-of-the-art, multi-megawatt wind turbines replacing older models in U.S. and Europe

2000
8,000 MW of electricity being generated from geothermal in 22 countries

2000
Renewable resources (including large hydro) contribute 9 percent of electricity in U.S. and over 19 percent globally

2000
Nuclear energy provides 20 percent of electricity in U.S. and almost 17 percent globally

2000
Fossil fuels (coal, oil, gas) provide 71 percent of electricity in U.S. and 64 percent globally

2000
99 percent of U.S. households have a color television

2001
Power outages and cost spikes, especially in California, spur growth of small solar and wind

2002
In Russia, the world's oldest nuclear plant closes down. Nuclear power still accounts for some 16 percent of world electricity and 20 percent in U.S.

2003
The largest power outage in U.S. history hits the northeastern states and part of Canada

2005
International Climate Change Task Force warns of a "runaway" greenhouse effect if global temperatures rise 3.6°F above preindustrial levels

2005
Kyoto Treaty reducing greenhouse gas emissions takes effect with wide participation — but without the U.S.

2005
Increase in development of run-of-river hydropower with "low-impact" hydro as the ultimate standard

ENERGY TIMELINE (continued)

2006
Wind turbines have grown in capacity to double and triple the size of earlier models; Germany and U.S. greatest users

2006
More states regulate greenhouse gas emissions; some 26 percent of the U.S. population affected

2008
Crude oil prices exceed $100 per barrel; U.S. gasoline prices exceed $4 per gallon

2009
Ocean energy resources projects proposed in the Pacific (by the U.S., Australia, New Zealand, Japan, and South Korea) and the Atlantic (by Canada, Portugal, France and the U.K.)

2009
Offshore wind energy production in Europe increases 54 percent in a single year

2010
11,000 MW of geothermal power generated worldwide and much more projected

2010
Solar photovoltaics production doubling every three years

2010
Hydrogen fuel cell technology used by several major U.S. companies

2010
An estimated 25 percent of the world's population (about 1.6 billion people) remain without access to electricity

Note: Suggestions for the Energy Timeline are always welcome. Please send them to energyforkeeps@aol.com.

GLOSSARY

A

acid precipitation (acid rain): any precipitation that primarily contains damaging sulfuric and nitric acids; may harm and/or destroy natural land or water habitats and corrode human structures including roads, buildings, and bridges

active solar: any system for collecting, storing, and releasing solar energy that requires an outside source of energy, such as fans or pumps, to operate system equipment

A.D.: any year after the birth of Jesus Christ; from the Latin, *anno Domini*, meaning "in the year of our Lord;" from 20 B.C. to 50 A.D. is 70 years

alloy: a mixture of different metals; for example, bronze, a mix of copper and tin; some alloys include metals mixed with non-metals (e.g., some kinds of steel are made of several metals plus carbon, a non-metal)

alternating current (AC): an electric current that reverses direction at regular intervals; caused by an alternating electromotive force (the force that produces an electric current)

alternative energy: sometimes defined as sources of energy other than fossil fuels, hydropower, or nuclear; often refers to transportation fuels other than gasoline — ethanol, biodiesel and hydrogen

alternator: an electric generator that produces alternating current

American Recovery and Reinvestment Act (ARRA): commonly referred to as the Stimulus or The Recovery Act; an economic stimulus package enacted by the 111th United States Congress in February 2009, intended to create jobs and promote investment and consumer spending during the recession

ampere (amp): a measure of the amount of current, or electrons, flowing in a wire over time; one ampere = 6.25 x 10^{18} electrons per second

anaerobic digestion: the breakdown of organic materials by bacteria in the absence of oxygen; results in the production of gases, primarily methane and carbon dioxide; occurs naturally or can be caused to occur under controlled conditions in an anaerobic digester

anemometer: a device for measuring wind speed

anode: the positively charged electrode in an electrical circuit or in an electrochemical reaction

aquafarming: the cultivation of fish and other water-dwelling organisms under controlled conditions

array: in general, a symmetrical arrangement of a large group, as in rows; in solar energy, usually refers to an arrangement of a large group of photovoltaic (solar) panels or mirrors

atom: the smallest particle of an element that retains the chemical properties of that element; composed of protons, neutrons, and electrons

B

balance of trade: the difference in value over a period of time between a country's imports and exports

barrage: an artificial obstruction, such as a dam or an irrigation channel, built in a river or other waterway to increase depth or divert flow

baseload power: the amount of power needed to supply the minimum anticipated demand for electricity at any given time

B.C.: any year before the birth of Jesus Christ; from 20 B.C. to 50 A.D. is 70 years

binary power plant: geothermal power plant that uses a heat exchanger to transfer heat to a second (binary means two) liquid that flashes to vapor and drives a turbine-generator

biomass: anything that is, or was once, alive: wood, plant, animal waste, and gases, such as landfill methane gas, ethanol, or other gases derived from them; a renewable energy resource

blackout: the loss of electricity, caused intentionally or by an electrical power failure

blast furnace: a furnace in which the combustion of a fuel is intensified using blasts of air or pure oxygen

brine: water containing large amounts of salts, particularly sodium chloride

GLOSSARY (continued)

brownout: a reduction in electric power; may be the result of a shortage or mechanical failure, or may be intentional in response to excessive consumer demand

by-product: something produced in the making of something else; a secondary product produced from the production of a primary product

C

capacity: in electricity generation, the manufacturer's rating of maximum electrical output for one or more generating devices

carbon cycle: the chemical cycle in which the element carbon naturally circulates in various forms throughout the living and nonliving systems of the earth over time

carbon footprint: a measure of our impact on the environment and, in particular, climate change; relates to amount of greenhouse gases produced in our day-to-day lives through burning fossil fuels for electricity, heating, and transportation

carbon monoxide: a gaseous molecule composed of one atom of carbon and one atom of oxygen; is highly toxic to animals and humans

carbon sink: components of the global ecosystem that store carbon; includes all plants, the ocean, old-growth forest floor litter (duff), soils, fossil fuels, and certain minerals such as limestone

carbon-based compound: element whose atomic structure causes it to join with a variety of other elements, forming the basis of many different compounds; the basis of all living things, as well as for fossil fuels (hydrocarbons) and many other substances, including diamonds and graphite

cathode: the negatively charged electrode in an electrical circuit or in an electrochemical reaction

Celsius (C): also centigrade; the temperature scale that registers the sea-level boiling point of water as 100° and the freezing point as 0°

central receiving tower: a concentrating solar power technology; a tall structure with a top section that contains a liquid, such as molten salt, water, or liquid metal, that has a high heat capacity; this liquid is heated by the reflection of solar energy from concentrating mirrors aimed at the tower's focal point

chain reaction: in physics, a method of releasing energy from the atom in a multistage nuclear reaction, in which the release of neutrons from the splitting of one atom leads to the splitting of others

central station power plant: a large plant that produces electricity for transmission and distribution to customers on a grid

charcoal: a material containing large quantities of carbon, formed by heating wood or other organic material in the absence of air

Clean Air Act (CAA): federal law designed to protect public health by setting standards and enforcement regulations governing air pollution emissions from energy production and other human activities

cogeneration: the process of doing work utilizing two forms of energy, usually thermal (heat) energy and electrical energy, both produced simultaneously from one source

coke: a fuel that burns very hot; used primarily in metal production; produced by removing mainly the sulfur, which makes iron brittle when smelted, from coal

combined cycle power plant: power plant in which two different turbines — most commonly a gas turbine accompanied by a steam turbine — work in succession to produce electricity; most gas-fired power plants are combined cycle plants

combustion: the process of burning, which is a chemical change requiring the presence of oxygen that results in the production of heat and light

Community Choice Aggregation: a program in which a local agency purchases power from a producer rather than from the local utility, while the utility provides the agency's customers with transmission, distribution and billing services

GLOSSARY (continued)

complete circuit: a complete and circular path for an electric current to follow as it moves through wires and electrical devices

compound: substance made of two or more elements that are bonded together chemically

concentrating solar power: any of the solar energy systems (solar dish engines, parabolic troughs, and central receiving towers) that focus, or concentrate, the energy of the sun for energy production or storage

condenser: a device that uses a cooling process to cause a vapor to condense to a liquid

conduction: the transmission of electric charge or heat through a conductor

conductor: in electricity, a substance or medium that conducts, or transmits, an electric charge; in thermal energy, a medium that allows the movement of heat through it

conservation: the controlled use and systematic protection of natural resources such as water, minerals, forests, and soil; also, the practice of avoiding and reducing waste, as in the production of or use of electricity

containment vessel: at a nuclear power plant, a large structure that houses the reactor core, its radiation shield, and the reactor core's maintenance equipment; the containment vessel is surrounded by an outer concrete building designed to prevent the escape of radiation in the event of an internal power plant accident or by an external event such as an airplane crash

control rod: in a nuclear power plant, a long rod made of material that absorbs neutrons; a number of these are inserted amidst the fuel rods in the reactor core; control rods are raised and lowered as needed to control the nuclear chain reaction, and thus the amount of heat energy produced

controller: in a wind turbine, a computerized device that receives information from all the sensors on the turbine (including anemometers, blade positions, temperatures, fault conditions, loads, vibration etc.) and uses this information to determine how to control all the various devices on the turbine

crude oil: unprocessed oil (petroleum) that varies in color and in thickness (viscosity); contains many different compounds, which can be separated and used for a variety of products, including energy fuels such as gasoline, heating oil, and butane

crust: in geology, the relatively thin, outermost rock layer of the earth

D

decompose: to become broken down into basic components or elements; to rot

deflect: to cause to turn aside

demand: in electrical power, the amount of electricity needed at any given time, based on the amount being used by all electrical devices connected to the power supply through the power grid

dense (density): the amount of mass, or matter, that is in a given volume of something; e.g., the molecules of a substance that is very dense are packed very closely together

deplete: to use up or consume

direct current (DC): an electric current that flows only in one direction

direct use geothermal: systems that use geothermal resources directly for heat energy rather than for producing electricity; includes space heating, greenhouse and fish farm operations, bathing and swimming at health spas, and industrial applications such as food and timber drying

disclosure: the act or process of revealing or uncovering; in energy management, information regarding which energy resources are being used to produce electricity by a power provider

distributed generation: supplying on-site electricity using small generating units; can be comprised of similar systems or a variety of different system types; distributed generation is used to manage peak loads, to add extra power for a region without having to build a large power plant, to provide electricity for remote locations, or for a vital industry (such as a hospital) that needs power even when grid power is unavailable

GLOSSARY (continued)

dry steam power plant: geothermal power plant that uses steam directly from a steam-filled geothermal reservoir

dynamo: an electric generator that usually produces direct current

E

ebb: to fall away or recede

ecological: pertaining to the science of the relationships between organisms and their environments

ecosystem: the community of all organisms living in an area and their interactions with the physical environment

electric current: the flow of charged particles through a conductive material

electrical energy: the energy of electrical charges, usually electrons in motion

electrochemical: the interaction of electrical and chemical phenomena

electrode: a solid electric conductor, such as a piece of metal, through which an electric current enters or leaves a solution containing an electrolyte; also, a collector or emitter of electric charge, such as found in a fuel cell

electrolysis: chemical reaction caused by passing an electric current through a liquid containing an electrolyte, resulting in the break down of the liquid's molecules; the electrolysis of water releases hydrogen and oxygen, for example

electrolyte: a chemical compound which, when molten or dissolved, usually in water, will conduct an electric current; an electrolyte solution

electromagnetic spectrum: radiated energy waves as described in terms of their wavelengths and frequencies, including gamma rays, X-rays, ultraviolet, visible light, infrared radiation, microwaves, radar, television, and radio; the sun is the largest natural source of electromagnetic radiation

electromagnetism: the study of the relationship between magnetism and electricity; the phenomena of producing electricity using magnetism and vice versa

electron: a negatively charged component of an atom; exists outside of and surrounding the atom's nucleus; can either be free or bound to a nucleus

element: the simplest possible chemical, made up of its own particular kind of atom; most elements occur naturally, though some have also been made artificially

encroach: to advance beyond usual or proper limits

energy carrier: a medium, system, or substance for conducting energy from its source to its users; examples are electricity and hydrogen

energy conservation: management of energy resources and energy use to prevent waste, reduce costs, and ensure future availability

energy conversion (transformation): the process of changing energy from one form to another

energy farm: a farm that grows plants specifically as biomass energy crops; sometimes refers to an array of generators for wind, solar, or ocean resources

enhanced geothermal system (EGS): an innovative geothermal technology in which water is pumped into artificially fractured hot rock to create a geothermal reservoir; also referred to as an "engineered" geothermal system

estuary: a coastal bay or inlet at the mouth of a river or stream, where large quantities of fresh water and saltwater mix together; if undisturbed, estuaries are very fertile and provide habitat for a variety of wildlife.

exempt: excused or released from a requirement

F

Fahrenheit (F): the temperature scale that registers the sea-level boiling point of water as 212°F and the freezing point as 32°F

GLOSSARY (continued)

fissionable: in nuclear power, an unstable element that is capable of being split; in a nuclear power plant, fissionable material — primarily one form of uranium (U-235) — is used to produce a nuclear chain reaction

fissure: in geology, an extensive crack, break, or fracture in rock

fixed-speed wind turbine: a wind turbine that always turns at the same speed, regardless of how fast the wind is blowing; the machinery of a fixed-speed wind turbine is simpler than that in a modern variable-speed turbine

flash power plant: a geothermal power plant that uses a process in which geothermal water is suddenly converted to steam to drive a turbine

fossil fuels: coal, oil, natural gas, and products made from them; fossil fuels are the remains of once-living (organic) plants and animals formed underground and subjected to intense heat and pressure over millions of years; have high concentrations of carbon and hydrogen and can be burned, producing energy (as well as polluting emissions)

fuel rod: at a nuclear power plant, pellets of uranium (U-235) that are arranged in long rods, which are collected together into bundles and placed in the reactor core

fumarole: steam and gas, venting from the earth's crust

G

gas turbine: power plant turbine that is driven by a continuous blast of hot gas from the combustion of natural gas combined with high-pressure air

gasification: the process of converting into or becoming a gas

generator: a machine that transforms (converts) mechanical energy into electrical energy

geothermal energy: the heat energy of the earth; the earth's natural heat emanating outward from its interior and constantly renewed from the radioactive decay of certain elements in the crust and other geologic processes; a renewable energy resource

geothermal reservoir (hydrothermal aquifer): a large volume of underground water saturating (filling) porous and permeable rock, superheated by the hot rock and nearby hot magma

global climate change: long-lasting changes in Earth's weather patterns and systems, resulting in dramatic, possibly harmful, changes in habitats and ecosystems worldwide; is thought by many researchers to be caused by the overall (global) warming of the planet, resulting from an excess of man-made greenhouse gases in the atmosphere

green energy: any energy source considered to be environmentally friendly; commonly associated with renewable energy sources, but also sometimes used when referring to nonrenewable sources that produce few pollutants

green pricing: offering customers the choice of paying additional fees on their utility bill in order to support production from new renewable energy facilities, either nearby or in another region or state

green waste: yard trimmings (usually leaves, grass clippings, and tree and bush trimmings), typically collected in specially designated containers and used for various purposes, including as a source of biomass energy

greenhouse effect: the trapping of heat energy from the sun in Earth's atmosphere, notably by water vapor and greenhouse gases such as carbon dioxide, nitrous oxide, and methane; the resulting heat energy warms the planet's surface

greenhouse gas: any gas in the atmosphere that contributes to the greenhouse effect

grid: the interconnected system that transmits and distributes electricity, including power plant(s), transmission and distribution lines, towers, substations, and transformers

groundwater: water that collects underground, mostly from surface water that has seeped down through cracks and pores in rock

H

habitat: the place that is natural for the life and growth of an organism

GLOSSARY (continued)

head: in hydropower, the distance that water falls before it hits a turbine generator

headrace: a channel that feeds water into a water wheel or (when generating electricity) a turbine

heat (thermal) energy: the energy that flows from one body to another because of a temperature difference between them; the effects of heat energy result from the motion of molecules

heat engine: any device that converts heat energy into mechanical energy; typical heat engines include steam engines, steam and gas turbines, internal combustion (vehicle) engines, and Stirling engines

heat exchanger: device used to transfer thermal (heat) energy from a liquid flowing on one side of a barrier to a liquid flowing on the other side

heliostat: an instrument in which a mirror is automatically moved so that it reflects sunlight in a constant direction

high and low tide: the rise and fall of the earth's oceans, caused mainly by gravitational forces of the moon and the sun

horsepower: originally the power exerted by a horse when pulling; now, a unit of power equal to 745.7 watts per minute

hot dry rock: a potential source of accessible heat energy within the earth's crust; a geothermal resource created when hot but impermeable (does not allow water to pass through) underground rock structures are fractured to allow infiltration of water, thus creating an artificial geothermal reservoir

hydrocarbon: any compound made up of hydrogen and carbon; will combine with oxygen when burned, producing heat energy; includes all the fossil fuels

hydrogen gas: colorless, combustible gas that can be used as an energy source; does not occur naturally by itself, and must be separated from another substance, such as from water, biomass, or a fossil fuel

hydrogen sulfide: a gas with a disagreeable odor, frequently dissolved in geothermal waters in small amounts; toxic at high concentrations

hydropower: the mechanical force of rapidly flowing or falling water from rivers or storage reservoirs; a renewable energy source

I

impoundment: a structure which allows the accumulation and storage of water in a reservoir; a dam placed across a river

incandescent light bulb: a glass bulb of inert gas (gas that is not readily reactive) that emits visible light as a result of passing electricity through a filament found inside the bulb, causing it to heat and glow

indirect (hidden) costs: the costs of producing a product (including electricity) that are not directly accounted for by an industry or utility, but may be borne by other sectors of society

Industrial Revolution: the shift to large-scale factory production brought about by the extensive use of machinery, often driven by coal-fired steam engines; generally thought to occur between the 1750s to the mid to late 1800s; resulted in dramatic social, environmental, and economic changes

industrial: the practice of making goods; often implies the production of large quantities of manufactured items, as found in factories

infrared: heat radiation; part of the electromagnetic spectrum radiated from the sun and other hot objects

internal combustion engine: an engine, used primarily in vehicles, in which fuel is burned within the engine itself, rather than fuel being burned in an external furnace (as in a steam engine)

J

jet stream: a narrow belt of westerly winds found at high altitude that can reach speeds of up to 230mph (370 km/h)

GLOSSARY (continued)

K

kilowatt (kW): 1,000 watts

kilowatt-hour (kWh): the energy expended when 1,000 watts of electrical power are used for one hour

L

liquefied natural gas (LNG): a petroleum fuel condensed from its normal gaseous form into liquid, a form more suitable to be transported

M

magma: hot, thick, molten rock found beneath the earth's surface; formed mainly in the mantle; when magma surfaces (usually from a volcano), it is called lava

magnetic field: a condition found in the region around a magnet or electric current where detectable magnetic force is found at every point and with distinguishable magnetic poles

mantle: the zone of the earth below the crust and above the core, primarily filled with a mixture of molten and solid rock

manufacture: to make a finished product, often using large-scale industrial operations

marine (ocean) current: movement of ocean water: two-way (tidal) or one-way (like the Gulf Stream)

mass: in physics, the measure of the quantity of matter that an object or body contains

mass-produce: to manufacture in large quantities, often using assembly lines

mechanical energy: the energy of an object as represented by its movement, position, or both

medieval: relating to a period in European history, usually between ancient cultures and the Renaissance (A.D. 476 to 1453), during which scientific and philosophical innovations were often suppressed

megawatt (MW): 1,000 kilowatts

methane gas: an odorless, colorless, combustible gas that can be used as an energy source; the primary component of natural gas and a source for hydrogen gas

microbe: a micro-organism; microscopic life form

micro-grid: a mini-version of an electric grid using distributed energy resources; can be operated as part of a utility's distribution system or "islanded" for separate use

micro-hydro: a hydroelectric plant of small size, not necessarily connected to a grid; most are "run-of-river" plants that do not require a dam

modular: designed with standardized equipment and dimensions to allow for flexible arrangement and the ability to add more units

module: in solar energy, a group of photovoltaic (solar) cells wired together into a single unit that can be grouped in any combination with other modules; in geothermal, a turbine-generator unit

mud pot: a type of hot spring containing boiling mud

multi-megawatt wind turbine: very tall wind turbine with huge blades that catch the faster wind speeds found higher from the ground; those most commonly used can generate between 1-3 megawatts of electricity; more advanced designs may produce up to 6 megawatts, with some designs as large as 20 megawatts

N

nacelle: in a wind turbine, a covered housing that protects the gear box, high- and low-speed shafts, generator, controller, and brake

NASA: the National Aeronautics and Space Administration; United States' space exploration agency; many scientific and technological advances that originated at NASA have been introduced into other industries

negative charge: one of two kinds of electric charge, the kind carried by an electron

GLOSSARY (continued)

net metering: a program offered by power producers that encourages grid-connected consumers to generate some or all of their own electricity using specific, usually renewable, resources; in many cases, this type of program allows the consumer's meter to turn backwards when they are producing more power than they are using, and some utilities will pay the consumer for the net excess power generated

neutron: an electrically neutral subatomic particle

nitric acid: a transparent, colorless to yellowish, corrosive substance; one of the components of acid precipitation

nitrogen oxides: gases formed mainly from nitrogen and oxygen; one of the damaging components of acid precipitation

nonrenewable energy: energy sources that do not regenerate themselves in a useful amount of time, including fossil fuels and nuclear fuels

nuclear fission: a reaction in which an atomic nucleus is split into fragments, releasing large quantities of energy; fission means "to split"

nuclear fuels: minerals, such as uranium, from which energy is liberated by a nuclear reaction

nuclear fusion: a reaction in which nuclei are combined (fused) to form a more massive nucleus, accompanied by the release of energy

nucleus: the positively charged central region of an atom (plural: nuclei)

O

ocean energy: the mechanical energy of ocean tides, currents, and waves, and the thermal energy of the solar and geothermal heat stored in waters of the ocean; a renewable energy source

ocean thermal energy: the solar and geothermal heat stored in waters of the ocean

Ocean Thermal Energy Conversion (OTEC): ocean energy technology that produces electricity — sometimes along with clean drinking water — by taking advantage of the temperature difference between warm surface ocean water and cold water from the ocean depths

oil refinery: factory where crude oil is separated into various components and cleaned to remove some impurities

oil rig: large collection of machines, hoists, and power equipment, established on land or on platforms or barges in open water; used to drill down into oil reserves found in underground rock

old-growth forest: forest having a mature ecosystem, including presence of old woody plants (mainly trees), and the wildlife and smaller plants associated with them; typically old-growth forest floors are made up of "duff," a rich layer of debris, decomposing matter, and leaves

one-way marine currents: deep oceanic currents, such as the Gulf Stream, that result from varying conditions of ocean water including differences in temperature and water density

organic decay: the breakdown of organic matter as a result of bacterial or fungal action; rot

organic: derived from living organisms

oscillating: to swing back and forth with a steady, uninterrupted rhythm

ozone: a highly reactive molecule made of three atoms of oxygen; high in the atmosphere ozone forms a protective layer that filters out harmful ultraviolet radiation; is formed at Earth's surface as a harmful component of photochemical smog

P Q

parabolic: a curved geometric shape based on the parabola; when radiant energy, such as sunlight, hits a parabolic surface and is reflected back, all the reflected radiant waves pass through one area of space in front of the parabolic surface known as the focus; in solar energy, parabolic surfaces, such as parabolic mirrors, are used to concentrate radiant waves from the sun

GLOSSARY (continued)

parabolic trough: a concentrating solar power technology that utilizes a long, trough-shaped parabolic reflector to focus the sun's energy onto a pipe that contains a liquid, usually an oil, that's used in a heat exchanger to boil water from steam

particulates: solid particles and liquid droplets suspended in the air, including smoke, soot, dust, and ash

passive solar: techniques using the structure of a building for heating or cooling that require no collectors, pumps, or other devices; examples include large, south-facing windows to allow solar energy in to warm the house, or awnings to block solar radiation to cool the house

peak load: the time(s) of day and times of year when consumers demand (use) the most electricity

peaking power: power to fill the electricity demand, or need, that exceeds the amount of baseload power available at any given time

penstock: a conduit or pipe that carries water from a storage reservoir or from upriver to a turbine

photochemical smog: a complex mixture of air pollutants, produced in the lower atmosphere by the reaction of hydrogen and nitrogen oxides when exposed to sunlight; is unsightly, damages vegetation, and leads to eye and respiratory ailments in animals and humans

photon: tiny bundles of electromagnetic radiation that move rapidly from one place to another at the speed of light; sometimes considered a flow of particles; the sun emits huge quantities of photons

photovoltaic: refers to the ability to convert photons into electrical energy; photons are used to dislodge electrons from atoms of silicon or other materials, causing them to migrate, producing an electric current

policy: a plan or general set of guidelines that reflects a particular set of values and influences specific actions and decisions

porous: in geology, able to hold water in spaces within rock

power: an application of force or energy; in physics, the rate at which work is performed or energy converted; technically, the amount of current times the voltage level at a given point measured in wattage or watts; often used interchangeably with the word "electricity"

positive charge: one of two kinds of electric charge, the kind carried by a proton

proton: a positively charged subatomic particle found in all nuclei

pumped storage: a system of generating electricity using water pumped from a lower reservoir to a higher storage site and later released to fall back to the lower reservoir when extra electricity is needed; used as a method of "storing" energy; generally, surplus electric power is used to pump the water uphill when electricity demand is low

R

radiant energy: energy transmitted in the form of rays, waves, or particles

radioactive: emitting radiation, either from unstable (fissionable) nuclei or from a nuclear chain reaction

reactive: an element or compound that tends to participate readily in chemical reaction

reactor core: in a nuclear power plant, the contained assembly of fuel rods, around which a liquid or gas flows in pipes to remove the resulting heat energy

rebate: return of a percentage of the cost of an item

regenerate: to renew the supply of something, such as an energy resource

renewable energy: any energy resource that is naturally regenerated or renewed within a useful amount of time and is to that extent inexhaustible

Renewable Portfolio Standards: a set of standards, adopted by a government, designed to ensure that a certain percentage of various renewable energy resources be included in the portfolio (or resource mix) of its power providers

GLOSSARY (continued)

resistance: in physics, opposition to the passage of electric current, causing electric energy to be transformed into heat

rotor: the rotating part (blades and hub) of an electrical or mechanical device such as a wind turbine

run-of-river (diversion): hydropower system that produces electricity while still maintaining the natural or near-natural flow of a river (as opposed to creating an impoundment to hold the river back to form a reservoir); most run-of-river systems divert some of the water to an electrical powerhouse and then return it to the river

S

scrubber: an apparatus used to remove impurities from gaseous emissions

semi-conductor: a material, usually silicon, which combines characteristics of a conductor (allowing current to flow) and an insulator (resisting current), suitable for use in electronics

silicon: one of the most abundant elements on Earth; always occurs in combination with other elements; high heat is required to isolate it; widely used in products such as glass, ceramics, computer microchips, and solar photovoltaic cells

sluice: an artificial channel for conducting water

smart grid: an electrical system that makes use of sensors and other devices to enhance the system's efficiency, as when supply is matched to demand in particular locations and at particular times

smelt: to melt ore (rock containing valuable minerals, especially metals) in order to separate the metal from the rock

solar cell: a photovoltaic device that converts solar energy into electrical energy using an electrochemical reaction in which electrons are caused to move, creating an electrical current

solar dish engine: a concentrating solar power technology that uses either one large, dish-shaped parabolic mirror, or a group of these mirrors, to concentrate the thermal (heat) energy of solar radiation onto a receiver; a heat engine in the receiver converts the concentrated heat into mechanical energy to drive an electrical generator

solar energy: heat and light radiated from the sun; a renewable energy source

solar panel: a group of around 10 solar photovoltaic modules (see solar cell) that are assembled together into a panel

spent fuel: fissionable material left over from a nuclear reaction; spent nuclear fuel is still radioactive, therefore toxic; classified as hazardous waste, and must be handled and stored properly for safety

stand-alone wind turbine: a wind turbine that is not part of a wind farm; most commonly used in remote or rural locations and often not connected to the electrical grid

static electricity: an accumulation of electric charge (as opposed to the movement of electric charge known as electric current); imbalance between positive and negative charges

steam reforming: a form of fuel processing often used to produce hydrogen gas, frequently from natural gas or biomass; uses a special process involving high-temperature steam and a catalyst (substance that increases the rate of a chemical reaction without being consumed in the process)

Stirling engine: an engine that has a sealed chamber where heat is focused on one side, causing the air inside to expand and push down on a piston; as the piston moves, air flows to the cold side of the engine where it is cooled; a second piston pushes the cooled air back to the hot side

strait: a narrow channel joining two larger bodies of water

subatomic particle: any of various units of matter below the size of an atom, including neutrons, protons, and electrons

GLOSSARY (continued)

substation: in electrical transmission, the location of the transformer equipment that decreases or increases the voltage of electric current flowing through transmission and distribution lines

sulfur oxides: pungent, colorless gases formed mainly by the combustion of fossil fuels; considered a major air pollutant

sulfuric acid: a colorless to dark brown, highly corrosive, dense liquid; sulfur oxide dissolved in water

sustainable: a process, system, or technology that does not deplete resources or cause environmental damage and thus lasts indefinitely; a school of thought that advocates preserving environmental, social, and economic resources, including energy resources, for future generations

synthetic: not natural; the combination (synthesis) of materials to form a product that may or may not occur naturally

system efficiency: input (of energy or work) versus output (of energy or work) of a system, often expressed as a ratio (energy in divided by energy out); theoretically, the ratio is never one-to-one

T

tailrace: the part below a water wheel or water turbine through which the used (spent) water flows

tectonic plates: the large sections of the earth's crust that are slowly moving over the mantle; the plates interact with one another at their boundaries, causing a variety of geologic phenomena including earthquake and volcanic activity

telegraph: apparatus historically used to communicate Morse code at a distance over a wire using electrical impulses

temperate zone: a region with a moderate climate, characterized by being neither too hot nor too cold

terrain: the surface features of an area of land

textile: cloth, especially that manufactured by weaving or knitting

thermal energy: see heat energy

thermal power plant: any type of power plant that uses heated fluid (normally steam) to drive one or more turbine-generators; usually a plan that burns fossil fuels, though geothermal and solar thermal plants require no combustion

thin film PV: a form of solar photovoltaic receptor which uses lightweight materials; a technology which makes solar panels more manageable and less costly than heavier forms of panel, but generally requires more surface area

thorium: a naturally occurring, slightly radioactive metal, estimated to be about three to four times more abundant than uranium in the Earth's crust

tidal currents: the two-directional, in and out and up and down movements of the ocean along coastlines

tidal fence: an ocean energy technology that uses a long, connected series of underwater turbines that tap tide currents to produce electricity

tidal power plant: marine current energy technology that uses the mechanical energy of ocean tides to produce electricity; traditional tidal systems situate turbines in a barrage (dam) through which the tides come in and out; newer designs use free-standing, generally submerged, turbines located at or near shorelines, bays, and channels

town gas: gas (composed mainly of hydrogen) that is manufactured from raw materials such as coal, coke, or oil; is distinguished from natural gas, which occurs naturally in underground deposits; during the 1800s town gas was widely distributed through pipelines to many cities and towns in Europe and America for light and heat

transformer: device used to "step-up" (increase) or "step-down" (decrease) the voltage of electric current

transmission lines: towers and wires through which high-voltage electricity travels over long distances

transmit: to send from one place to another

turbine: bladed, wheel-like device caused to spin by the force of pressurized steam or gas, wind, or moving water; used in electricity production to drive an electrical generator

GLOSSARY (continued)

U

U.S. Environmental Protection Agency (EPA): a federal agency of the United States with the mission of protecting the nation's natural environment; establishes and enforces regulations through a network of regional offices

ultraviolet: radiant waves that are part of the electromagnetic spectrum; are invisible to the human eye; solar ultraviolet radiation comes in several wavelengths, one of which is harmful to biological life, but most of which is absorbed by upper atmospheric ozone layer

unburned hydrocarbons: air pollutants that come from the incomplete combustion of fossil fuels and from the evaporation of petroleum fuels, industrial solvents, painting and dry cleaning chemicals

uranium: a heavy, silvery-white metallic element that is radioactive and toxic; exists in 14 different forms, or isotopes; is extracted from ores for use in research, nuclear fuels, and nuclear weapons

V

vaporize: to convert into a vapor, the gaseous state of a substance

variable-speed wind turbine: turbine that can respond to wind speed changes

voltage: the measure of the electrical force that "pushes," or drives, an electric current

W X Y Z

wastewater: the collective discharge from toilets, sinks, showers, washing machines, storm-sewers, etc.; can be cleaned, or "treated," to remove most of the toxic components and then used for purposes other than consumption by animals or humans

water cycle: the natural process of the movement of Earth's water as it evaporates from bodies of water, condenses, precipitates (rains, sleets, hails, snows) and returns to those bodies of water, in a continuous cycle

watt (W): the rate of electrical current flow, when one ampere is driven, or "pushed," by one volt

watt-hour (Wh): the energy expended when one watt of electrical power is used for an hour

wet-cell battery: a battery, or "cell," in which an electrochemical reaction occurs in an electrolyte

wetland: a lowland area, such as a marsh, swamp, or estuary, that is saturated with moisture; provides a rich habitat for wildlife; absorbs heavy metals and filters out toxins, releases oxygen into the air while removing carbon dioxide and other greenhouse gases; provides flood control and is a significant factor in the recharge of groundwater

wind energy: the mechanical force of moving air saturated with kinetic energy; a renewable energy source

wind farm: a cluster of wind turbines located in areas with reliably favorable wind speeds, such as on high windy mountain passes or gusty open plains; can also be situated on farms or ranches alongside crops and grazing animals

ADDITIONAL INFORMATION RESOURCES

These listings are selected from the wealth of information available on all aspects of energy use. First listed are information resources specific to each chapter, followed by a more general information section. Many of the listings include great website links.

CHAPTER 1: ENERGY HISTORY

California Energy Commission
Energy Time Machine
www.energyquest.ca.gov/time machine

Extensive timeline of energy history from the dawn of history to present day.

Milestones in the History of Energy and Its Uses
EIA Energy Ant Kids Site
www.eia.doe.gov/kids/milestones

Traces significant events in the history of energy; links to "Pioneers in Energy" and "Energy in the United States, 1635-2000."

See also "Other Information Resources," pages 181-184.

CHAPTER 2: ELECTRICITY

Electricity Online
ThinkQuest
www.thinkquest.org

Explores the physics, practical applications, and history of electricity in an interactive, online format.

See also "Other Information Resources," pages 181-184.

CHAPTER 3: BIOMASS

California Biomass Energy Alliance
805-386-4343
www.calbiomass.org

General biomass information; specific information on California biomass power plants; ask an expert.

National Renewable Energy Laboratory (NREL)
Clean Energy Basics
About Biomass Energy
www.nrel.gov/learning/re_biomass.html

Information about state-of-the-art biomass technologies; general information on using biomass for energy.

U.S. Department of Energy (DOE)
Office of Energy Efficiency and Renewable Energy
Biopower Division
www1.eere.energy.gov/biomass/abcs_biopower.html

Information on all aspects of using biomass for energy; links to related organizations and information sources; library; photo gallery.

CHAPTER 3: GEOTHERMAL

GeoHeat Center
Oregon Institute of Technology
541-885-1750
http://geoheat.oit.edu

General information on geothermal energy, especially its use at lower temperatures; geothermal heat pumps; where geothermal resources are located and being used; access to experts; links to other information sources.

Geothermal Education Office
www.geothermal.marin.org

Information and educational materials on all aspects of geothermal energy; geothermal curriculum unit, videos, maps; posters.

Geothermal Energy Association
www.geo-energy.org
202-454-5261

Trade association and advocacy organization promoting use of geothermal energy; lists of all U.S. geothermal power plants; papers on environmental and economic impacts of geothermal energy development.

ADDITIONAL INFORMATION RESOURCES (continued)

Geothermal Resources Council
www.geothermal.org
530-758-2360

Primary membership and educational organization for the geothermal industry worldwide; library of downloadable semi-technical papers on geothermal energy.

U.S. Department of Energy (DOE)
Office of Energy Efficiency and Renewable Energy
Geothermal Technologies Program
www1.eere.energy.gov/geothermal

Information on all aspects of geothermal energy and links to other geothermal energy sites.

CHAPTER 3: HYDROPOWER

Bonneville Power Administration (BPA)
800-282-3713
www.bpa.gov
General information on large hydro and related subjects.

Bureau of Reclamation Power Program
Hydropower Information
www.usbr.gov/power

Topics covered include history of hydropower in the United States; background information on hydropower, major hydropower producers; links to other sources of information; educational materials for K-8, including "Nature of Water Power."

Foundation for Water and Energy Education
800-279-6375
www.fwee.org

Many educational materials on hydropower; information on all aspects of hydropower including environmental impacts.

National Hydropower Association
202-682-1700
www.hydro.org

Advocacy organization promoting the widespread use of hydropower; access to basic hydropower information.

U.S. Department of Energy (DOE)
Office of Energy Efficiency and Renewable Energy
Hydropower Division
www1.eere.energy.gov/windandhydro

Information on all aspects of hydropower; links to other hydropower resources and organizations.

CHAPTER 3: OCEAN

Ocean Energy Council
561-793-0320
www.oceanenergycouncil.com

Basic information on ocean energy and extensive links to government and industry sites.

Ocean News & Technology
www.oceanrenewable.com

Information about conferences, legislation, and jobs in the marine renewable industries.

U.S. Department of Energy (DOE)
Office of Energy Efficiency and Renewable Energy
Ocean Topics
www.eere.energy.gov

Information on all aspects of ocean energy and links to other ocean energy sites.

CHAPTER 3: SOLAR

American Solar Energy Society
303-443-3130
www.ases.org

Advocacy organization promoting widespread use of solar energy; information on all aspects of solar energy; magazine: *Solar Today*; Solar Guide Fact Base; publications; educational materials: videos, slides, activities.

Florida Solar Energy Center Teacher Resources
www.fsec.ucf.edu/ed/teachers

Information on all aspects of solar energy; student contests such as Junior Solar Sprint and Hydrogen Sprint; offers many teaching resources including units on energy in general, solar energy, alternative fuels, and environmental issues.

ADDITIONAL INFORMATION RESOURCES (continued)

Project Sol
Arizona Public Service (APS)
www.aps.com/my_community/solar/solar_6.html

A solar education site developed by APS (an Arizona power supplier); topics include energy from the sun, electrical energy, inside PV systems, power for the future; solar data; virtual tour of a photovoltaic cell.

U.S. Department of Energy (DOE)
Solar Decathlon
www.solardecathlon.gov/for_teachers.cfm

Interactive website for teachers covering various topics, including solar energy and energy efficiency.

CHAPTER 3: WIND

American Wind Energy Association
202-383-2500
www.awea.org

Advocacy organization promoting widespread use of wind energy; information on all aspects of wind energy; online bookstore; "Wind Energy Weekly" covers wind industry, global climate change, and energy policy; resource library; information on specific wind energy projects.

U.S. Department of Energy (DOE)
Office of Energy Efficiency and Renewable Energy
Wind Energy Program
www1.eere.energy.gov/windandhydro

Information on wind energy basics, including how wind turbines work; wind turbine research, and wind energy projects; links to other organizations; resources for teachers and students; photo gallery.

Wind Energy Resource Atlas of the United States
National Renewable Energy Laboratory
http://rredc.nrel.gov/wind/pubs/atlas

Atlas showing the quality of wind energy resources in various parts of the United States.

CHAPTER 3: HYDROGEN

Fuel Cell Store
303-237-3834
www.fuelcellstore.com

Fuel cell products for classroom and for the general public; products include fuel cell demonstration kits, fuel cell systems and accessories; resources for students and teachers, including fuel cell experiments, books, posters, and videos.

National Hydrogen Association
202-223-5547
www.hydrogenus.org

Advocacy organization promoting the widespread use of hydrogen fuel; basic information on hydrogen fuel; resources for students and educators.

Schatz Energy Research Center
707-826-4345
www.humboldt.edu/~serc

Working in affiliation with Humboldt State University's Environmental Resources Engineering program, develops and promotes renewable energy technologies, especially hydrogen fuel cells, zero emission vehicles, and solar hydrogen power systems; information on all aspects of hydrogen and fuel cells; educational materials.

CHAPTER 3: FOSSIL FUELS

Petroleum Education
Paleontological Research Institution
607-273-6623
www.museumoftheearth.org

"From the Ground Up: The World of Oil" covers all aspects of oil including geology basics, oil history, hydrocarbon systems, daily uses of oil.

U.S. Department of Energy (DOE)
Fossil Energy Division
www.fe.doe.gov

Extensive information on all aspects of fossil fuel production and use in the United States and globally; recent fossil fuel news items; clean coal and natural gas technologies; "For Students" section.

ADDITIONAL INFORMATION RESOURCES (continued)

CHAPTER 3: NUCLEAR

Nuclear Energy Institute
202-739-8000
www.nei.org

Advocacy organization promoting the use of nuclear energy; information on nuclear technologies; public policy issues; nuclear data; library; "NEI Science Club," teachers and kids site that includes games, information, curricular materials.

U.S. Department of Energy (DOE)
Office of Nuclear Energy, Science and Technology
www.ne.doe.gov

Information on all aspects of nuclear energy; nuclear power research; space and defense power programs; nuclear facilities management; nuclear fuel supply security; public information.

CHAPTER 4:
ENERGY, HEALTH, AND THE ENVIRONMENT

Earth Island Institute
415-788-3666
www.earthisland.org

Institute researching and promoting a wide variety of projects on conservation, preservation, and restoration both nationally and globally; "Earth Island Journal," many publications; news and citizen action alerts; information on starting your own action project.

National Oceanic and Atmospheric Administration
202-482-6090
www.noaa.gov

Researches and disseminates information on all aspects of climate, weather, and the oceans; weather forecasting satellite imagery; ocean exploration; fisheries; climate research; air quality; coastal services; undersea laboratory; library and archives; photo library.

See also Chapter 5 information resources and "Other Information Resources" pages 181-184.

CHAPTER 5:
ENERGY MANAGEMENT STRATEGIES AND ENERGY POLICY

Alliance to Save Energy
202-857-0666
www.ase.org

Advocacy organization promoting energy efficiency; energy efficiency programs, including "Energy Science Fair," "Green Schools," "New School Construction," and "Downloadable Educator Lesson Plans."

American Council for an Energy-efficient Economy
202-429-2248
www.aceee.org

Organization dedicated to advancing energy efficiency; advises on and provides educational information on energy policy, energy efficient buildings, industry, transportation; publications and other consumer information. Look for "Consumer Guide to Home Energy Savings."

Astronomy Picture of the Day (APOD)
National Aeronautics Space Administration
http://antwrp.gsfc.nasa.gov/apod/ap001127.html

Satellite composite photo taken Nov. 27, 2000, shows "Earth at Night": highlights developed or populated areas of the earth's surface; can be used to demonstrate differences in resource consumption between developed and developing nations.

Carbon Footprint
www.carbonfootprint.com/calculator.aspx

An educational and easy-to-use carbon footprint calculator.

Redefining Progress: Sustainability Program
510-444-3041
www.rprogress.org/programs/ecological.footprint

Partnership of organizations dedicated to sustainability; calculate your own ecological footprint; ecological footprint concepts and methods; sustainability education resources; publications.

ADDITIONAL INFORMATION RESOURCES (continued)

Rocky Mountain Institute
970-927-3851
www.rmi.org

Fosters sustainable social, economic, and environmental practices; information on energy, climate, water, transportation, energy efficient buildings; Kids site; educational materials; newsletter, bookstore.

Union of Concerned Scientists
See page 184.

U.S. Environmental Protection Agency (EPA)
www.epa.gov

Federal government health and environment regulatory agency; information on many topics including laws and regulations, environmental management, health topics, pollution prevention, economics, compliance and enforcement; educational resources; extensive Global Warming Site, including Kids site and educator materials and information.

Worldwatch Institute Resource Center
202-452-1999
www.worldwatch.org

Independent research organization advocating environmental sustainability; resource center topics include energy resources, climate change, transportation pollution, biodiversity, food, population, and water issues; publications and news alerts.

See below for more information on sustainability, energy policy, and management.

OTHER INFORMATION RESOURCES

American Council on Renewable Energy (ACORE)
202-293-1123
www.acore.org

Non-profit organization formed to accelerate the adoption of renewable energy technologies into the mainstream of American society; focus on trade, finances, and policy; promotes all renewable energy options.

Acorn Naturalists
800-422-8886
www.acornnaturalists.com

Books and other teaching materials on many topics including environmental education, outdoor education, science inquiry, interpreting cultural and natural resources, "GEMS" ("Great Explorations in Math and Science"), earth science, ecology, plant and animal studies, and the ocean.

Ask an Energy Expert
1-800-DOE-3732
www.eere.energy.gov/industry/bestpractices/energymatters/archives.html

A division of U.S. DOE Office of Energy Efficiency and Renewable Energy; answers questions ranging from how to make your school more energy efficient to specifics on the use of renewable energy.

Bonneville Power Administration (BPA)
800-282-3713
www.bpa.gov

General information on water, hydroelectricity, energy conservation, electric safety, resource planning and BPA history.

California Energy Commission (CEC)
General: 916-654-4287
Toll Free in California: 1-800-555-7794
www.energy.ca.gov

Consumer Energy Center Website; information about energy efficiency, energy statistics, and renewable energy; rebate information news releases; programs include energy efficiency, renewable energy development, alternative fuel vehicles.

California Energy Commission Kids Site: Energy Quest
www.energyquest.ca.gov

"Time Machine," "The Energy Story" (all aspects of energy and energy resources), games, energy terms, "How Things Work," science projects, "Ask Professor Questor," teacher and parent resources.

ADDITIONAL INFORMATION RESOURCES (continued)

California Mineral Education Foundation
916-655-1050
www.calmineraled.org

Charitable education corporation developed to provide mineral education programs for K-12 teachers. Covers wide variety of geological topics, as well as mining and processing of minerals.

Center for Energy Efficiency and Renewable Technologies (CEERT)
916-442-7785
www.ceert.org

Based in Sacramento, public interest coalition working towards policy change and public education regarding the use of sustainable, environmentally sound methods to meet California's energy needs. Up-to-date information on renewable energy technologies, energy efficiency, and energy policy.

Chelsea Green: Books for Sustainable Living
800-639-4099
www.chelseagreen.com

Wide range of sustainable living books and some videos on topics such as energy-efficient homes, stand-alone renewable energy systems, ecological architectural design, and renewable energy.

Energy for Keeps
www.energyforkeeps.org

Supplemental information for this book, including student activities, information for teachers, reader comments and suggestions.

Energy Ant: DOE Kids Zone
www.eia.doe.gov/kids

Energy history, articles on various energy topics, "What is Energy," "Kids Corner," "Energy Quiz," teacher resources.

Foundation for Water and Energy Education
www.fwee.org

Information related to the use of water as a renewable resource in the Pacific Northwest.

The Franklin Institute
214-448-1200
www.fi.edu

Museum online resource; science history; energy information; online study unit topics include wind, plate tectonics, oceans; links to many other resources; "Community Science Action Guides" include global warming, fossil fuel depletion, nuclear energy, energy resources, and visual animations of energy at work.

How Stuff Works
www.howstuffworks.com

Reliable information source on just about every topic, including many specific energy-related topics.

Interstate Renewable Energy Council (IREC)
518-458-6059
www.irecusa.org

Non-profit organization formed to accelerate the sustainable utilization of renewable energy sources and technologies in and through state and local governments and community activities; strong education and community outreach programs.

National Energy Education Development (NEED)
703-257-1117
www.need.org

Partner with U.S. DOE's Rebuild America and "Energy Smart Schools." Information about energy resources, including how their use impacts the environment; K-12 curriculum material including hands-on activities about the science of energy, electricity, efficiency and conservation; training and professional development; photo gallery.

National Energy Foundation
801-908-5800
www.nef1.org

Information about renewable energy, efficiency, and conservation. Materials catalog, NEF Academy for professional development, Energy Action Programs (energy awareness and energy management for schools, community, home), student programs include "Academy of Energy," "Fueling the Future," and "Igniting Creative Energy."

ADDITIONAL INFORMATION RESOURCES (continued)

National Renewable Energy Laboratory (NREL)
303-275-3000
www.nrel.gov/education

U.S. DOE's laboratory for renewable energy and energy efficiency research and development; general information on state-of-the-art renewable energy technologies; Office of Education Program provides renewable energy and energy efficiency curriculum, activities, projects; student competitions; teacher training, including direct access to current renewable energy research.

National Science Resources Center
Smithsonian Institution/The National Academies
www.si.edu/nsrc

Many educational resources on all topics, including energy; publications; science newsletter; links to many resources; science curriculum units for both middle school and K-6.

Northeast Sustainable Energy Association
413-774-6051
www.nesea.org

Education section provides interdisciplinary K-12 materials on energy, transportation, and the environment; links to Green Car Club, Clean Energy, and Green Buildings.

NOVA Science in the News
Australian Academy of Science
www.science.org

Up-to-date linked information on various science topics, geared for high school level; categories include environment, physical sciences, and technology; includes links to such topics as climate, electromagnetism, and plate tectonics.

Renewable Energy Partnership
www.repartners.org

Helps public power, electric co-ops, and tribal utilities get current, reliable information about renewable power; a one-stop shop for researching renewable energy options; information on best green power marketing practices; deep-links to topics such as industry calendars, green power, state and federal funding opportunities, case studies, transmission studies, tools for market research, tools for identifying and screening renewable resources; resources to educate customers.

Renewable Energy Policy Project (REPP)
www.crest.org

Information on renewable energy; energy and the environment, efficiency, and policy issues; library archives; "Global Energy Marketplace," e-mail newsletter; up to date news; recent trends.

Renewable Energy Project Kits
Pembina Institute, Canada
www.re-energy.ca

Provides background information on selected renewable energy resources (including wind, hydropower, solar, biomass); includes detailed directions for building working models.

Renewable Energy World
www.jxj.com/magsandj/rew

Website containing many articles from magazine of same title; global coverage of state-of-the-art renewable energy projects and policy issues; information is rather technical, but students can skim for general information; one of the best sources for up-to-date information; check to see if it will give you free subscription to the print-version magazine.

Sustainable Energy Coalition
202-293-2898
www.sustainableenergycoalition.org

Advocacy organization that promotes federal support for energy efficiency and renewable energy technologies; energy facts and statistics; energy policy information; links to many energy experts.

ADDITIONAL INFORMATION RESOURCES (continued)

Tennessee Valley Authority Kids Site
www.tvakids.com

Information on protecting the environment, making electricity, "Green Power," electrical safety, TVA history; teacher resources include a K-12 renewable energy curriculum and "Energy Sourcebooks" with teacher guides and energy education activities.

Union of Concerned Scientists
National Headquarters
Phone: 617-547-5552
West Coast Office
Phone: 510-843-1872
www.ucsusa.org

Partnership of scientists and citizens for scientific analysis, policy development and citizen advocacy promoting practical and sustainable environmental solutions in many areas including energy use and pollution; programs include support for renewable energy development and policies.

U.S. Department of Energy (DOE)
Energy Information Administration
202-586-8800
www.eia.doe.gov

Ask an Expert; Energy data, analyses, forecasts, and publications about specific energy resources, as well as general publications such as "Monthly Energy Review," the "Annual Energy Review," the "Short-Term Energy Outlook," and the "Annual Energy Outlook."

U.S. Department of Energy (DOE)
Office of Energy Efficiency and Renewable Energy
202-586-9220
www.eere.energy.gov

Portals to related U.S. DOE offices, as well as to many other programs related to energy efficiency and renewable energy; energy education programs include energy curriculum, science projects and activities, student competitions, and student resources; oversees "EnergySmart Schools" and "Rebuild America" programs.

INDEX

i = illustration
t = table

A

AC. *See* Alternating current
Acid precipitation, 136
Acid rain. *See* Acid precipitation
Active solar, 84, 148. *See also* Solar energy
Agricultural waste, 39
Air conditioning, 144
    peak load and, 145,152
Air pollution, 123, 135t, 136, 142. *See also* Pollution
    Clean Air Act (CAA) and, 151
    cost of, 151, 153
    solutions, 138–139
    sources of, 134i, 135t, 136
Air Quality Management District, 151
Alaska, hot springs in, 54
Altamont wind farm, 98, 99i
Alternating current (AC), 23, 24
Alternative energy, 35
American Lung Association, 136
American Recovery and Reinvestment Act (ARRA), 155
Ampere, Andre Marie, 30
Amperes, 30
Anacapa Island, 85
Anaerobic digestion, 44, 109
Anemometer, 96, 96i
Anode, 111
ARRA, 155
Atoms, 30, 125–126, 127
Automated Demand Response (Auto DR), 153
Automobiles
    fuel cells in, 108, 110, 112, 147
    invention of, 23
    proliferation of, 23, 25

B

Bain, Addison, 110
Baseload power, 32
    biomass and, 46
    coal and, 124
    geothermal and, 58
    hydropower and, 69
    natural gas and, 124
    nuclear and, 130
    oil and, 124
    solar and, 92
    tidal and, 80
    wind and, 101
Batteries, 30, 31
    solar, 85, 86, 88, 148

wet cell, 22
wind and, 97, 101
Benz, Karl, 23
Big Creek, 63
Biodiesel, 35
Biofuels, 40–41, 44
Biogases, 40–41, 44
Biomass, 33, 34i, 39–46, 47–48
    biofuels and, 40–41
    biogases and, 40–41, 43
    by-products, 45
    definition of, 39
    electricity, 43–44, 45, 46, 47–48
    as energy source, 34i, 41
    in hydrogen production, 110, 111
    power plants and, 43–44
    solid state of, 39
    volume of, 43–44
Biomass power plants
    biofuels and, 40–41
    capacity of, 43–44
    carbon dioxide (CO_2) and, 44, 44i
    fast-growing trees and, 39i, 41, 41i
    large, 43, 43i
    proliferation of, 45
    small, 44
Birds, wind turbines and, 99, 101
Bishop Creek Hydropower Project, 67
Blackouts, 32, 147, 152
Blast furnaces, 20, 20i
Brownouts, 32, 147, 152
Buffalo Ridge, 97
Building efficiency, 144–145, 145i

C

CAA. *See* Clean Air Act
California Fuel Cell Partnership, 112
Carbon cycle, 141
Carbon dioxide (CO_2), 135t, 136, 142
    air pollution and, 133, 134i, 135t, 136, 142, 143
    biomass and, 44, 44i, 109
    in carbon cycle, 141
    climate change, 137, 139, 141, 142
    from fossil fuels, 118, 134i, 135t
Carbon footprint, 138, 155
Carbon monoxide (CO), 109, 134i, 135t, 136
Carbon sinks, 141, 142
Carbon, storing, 142
Carriers, energy, 30, 34i, 108
Cast iron, 20
Cathode, 111, 111i

Center for Resource Solutions, 154
Central receiving towers, 88, 90, 90i
CFL (Compact fluorescent light bulb), 144, 146
Channel Islands National Park, 85
Chena Hot Springs, 54
CHP. *See* Combined heat and power
Circuits, electrical, 30
Cities
    industrialized, 19, 21i, 24
    wind turbines in, 100
Clean Air Act (CAA), 151
Clean energy, 35, 133, 138
    climate change and, 139
    solutions, 138, 142
Climate, 130, 133
    changes, 130, 135t, 138–139, 140
    effects of changes on, 140
    greenhouse effect and, 135t, 137, 137i
    solutions, 141–142
CO_2. *See* Carbon dioxide
Coal, 18–19, 117
    cleaning, 121
    coke and, 19, 20, 20i
    convenience of, 24–25
    demand for, 19, 20
    electricity from, 120, 120i
    gasification, 121
    mining, 18, 18i, 19, 42, 112, 117, 118, 120, 120i, 124, 146
    pollution and, 118, 123, 135t, 136
    power and, 18i
    power plants, 120, 120i, 121
    in steam engines, 18, 18i, 19–21, 23
    sulfur in, 20, 121, 135t
Cofiring, 44
Cogeneration, 41, 121, 122, 149
Coke, 19, 20, 20i
Combined heat and power (CHP), 121, 122, 149
Community choice aggregation (CCA), 154
Compact fluorescent light bulb (CFL), 144, 146
Concentrating Solar Power (CSP), 88
    parabolic troughs and, 89, 89i
    size of, 92
    solar dish engines and, 88, 88i, 92
    towers and, 88, 90, 90i
Condensers, 43, 43i
Conductors, 22

INDEX (continued)

Congress (U.S.), 151
Conservation
    energy, 26, 143, 144
    home, 144, 145, 145i
Colorado River, 65
Copper wire, 22
Coral reefs, 140
Cotton mills, 19i
Credits, tax, 155
Crude oil, 118-119, 119i. *See also* Oil
CSP systems. *See* Concentrating
    Solar Power systems
Current, 22, 23, 24, 28

D
Daimler, Gottlieb, 23
Dams, 63, 64, 65, 69, 70
Demand Response systems, 153
Deregulation, 150
Dinosaurs, 117
Direct current (DC), 23, 24
Distributed Generation (DG), 147
Dioxin, 45
Dufay, Charles, 22
Dynamos, 22, 23

E
Earth, crust of, 50, 50i
Earthquakes, 50, 56, 58
Ecological footprint, 138, 155
Edison, Thomas, 23
EGS. *See* Enhanced geothermal
    systems
Einstein, Albert, 24
    photoelectric effect and, 86
Electric alternator, 24
Electricity
    baseload power and, 32, 46, 58,
      69, 80, 92, 101, 124, 130
    biomass and, 26, 41, 42-46
    blackouts and, 32, 147, 152
    conservation of, 143-145, 145i, 146
    definition of, 30
    development of, 23-24
    electromagnetism and, 22, 28
    energy sources for, 35-36
    experiments with, 22
    fossil fuels and, 117-119, 120-122,
      123
    Franklin, Benjamin and, vii-viii, 22
    generation, 27, 28, 34i, 61
    generators, 23-24, 28, 29, 29i
    hidden costs of, 153

    hydropower and, 25, 61-70
    meters, 31, 32, 152
    nuclear energy and, 127-129, 130,
      131
    power grids and, 147
    pricing, 32, 152, 153
    solar energy and, 25, 25i, 26, 83-
      92
    storage of, 31
    transmission, 31, 24
    turbines and, 25, 27, 27i, 28-29,
      29i, 96, 96i, 97
    usage, 33, 35, 143, 144
    wind power and, 26, 95-101
Electrolysis, 22, 92
    nonrenewable, 110
    renewable, 108-109, 109i, 114
Electrolytes, 108, 111i
Electromagnetism, 22
    in electricity generation, 28
Electronic communication, 22
Electrons, 30
Emissions. *See also* Air pollution
    geothermal power plant, 57, 57i
Energy, 17, 28. *See also* specific energy
    types
    carriers, 30, 34i, 108
    chain, 28
    choices, 35
    management of, 152-155
Energy conservation, 26, 143, 144
    in home, 144, 145, 145i
Energy crops, 41, 44, 39
Energy efficiency, 145, 145i, 146,
    146i, 149
    in buildings, 143, 144, 145, 145i,
      146
Energy farms, 41, 76, 78, 98
Energy policies, 150-151
Energy resources, 17-18, 33-36
    for generating electricity, 34i, 36i
    renewable v. nonrenewable, 33-34i
ENERGY STAR program, 153
Energy timeline, 159-164
Enhanced geothermal systems
    (EGS), 56
Environmental policy, 150, 151
EPA. *See* U.S. Environmental
    Protection Agency
Erosion, 41
Ethanol, 35, 39
Evaporation, 62, 62i

F
Factories, 19, 61, 117, 119, 153
    expansion of, 20
Falls Creek, 68
Faraday, Michael, 22, 108
Feed-in tariffs, 155
Fermi, Enrico, 26
Fire, 17, 39. *See also* Wood
Fish
    hydropower plants and, 65, 69
    ladders, 65, 65i
    run-of-river hydropower and, 66
Footprint. *See* Carbon Footprint
Ford, Henry, 24
Forests, 39, 40. *See also* Rainforests
    protecting, 142
    thinning, 45
Fossil fuel(s), 18-19, 22-23, 25, 26,
    117-124
    air pollution and, 118, 121, 123,
      133-136
    coal and, 117, 118, 120-121, 123
    conserving, 26, 144
    cycle, 134i
    dinosaurs and, 117
    dominance of, 18, 19, 25, 117
    electricity from, 26, 35, 117-119,
      120-122, 123
    in electrolysis, 110
    as energy source, 22-23, 26, 33-35
    formation of, 117-118
    history of, 117
    hydrogen in, 107, 112, 114, 117,
      118
    increased usage of, 117
    natural gas and, 117, 119, 122,
      123
    oil and, 18-19, 117, 118-119, 122,
      123
    overuse of, 26
    pollutants, 118, 121, 123
    portability of, 119, 123
    shortages of, 26, 117, 122, 123
Franklin, Benjamin, vii-viii, 22
Fuel cells, 26, 107, 110-111, 111i,
    112, 113
    automobiles and, 112, 114, 147
    hydrogen, 26, 107, 110-111, 111i,
      113
    usage of, 112, 112i, 113, 114
Fuel rods, 128, 131
Fumaroles, 49, 51, 51i, 55

INDEX (continued)

G

Garbage, as biomass, 39, 42, 45, 46
Gas turbines, 121
Gasification, 40–41, 110
    coal, 121
    in hydrogen production, 109, 121
Gasoline, 23, 25, 26, 35, 119. See
      also Oil
    efficiency of, 146
    v. hydrogen, 108, 113
    usage, 25, 118
Generators, 23–24, 28, 29, 29i
Geothermal energy, 33, 34i, 49–58,
      59–60
    developments in, 26
    direct use of, 149, 149i
    electricity from, 52–53, 59–60
    as energy source, 33, 34i
    minerals and, 58
    power plants, 28, 29i, 49
    usage of, 17, 49, 57, 57i
Geothermal heat pumps, 149, 149i
Geothermal locations, 55, 56, 56i
Geothermal power plants, 26, 52, 55,
      55i, 57, 58
    binary, 53, 53i, 54
    capacity of, 58
    development of, 55
    dry steam, 53, 57
    emissions, 57
    flash steam, 52i, 55
    groundwater and, 57
    hot dry rock, 56
    location of, 49, 55, 57
    size of, 54
    types of, 52–54, 54
Geothermal reservoirs, 51, 51i, 56,
      57, 58
    geothermal power plants and, 52i
    minerals in, 57, 58
    replenishing, 52, 52i, 55
Geysers, 49, 51
The Geysers, 55, 55i
Glen Canyon National Recreation
    Area, 86
Global warming, 135t, 138–139
    effects of, 135t, 140
    solutions, 141–142
Gramme, Zenobe, 23
Grand Coulee Dam, 64
Gray, Stephen, 22
Green energy, 35
Green pricing, 154

Green Tags (TRECs), 154
Green waste, 39, 42, 44, 45, 46
Green-E, 154
Greenhouse effect, 135t, 137, 137i
Greenhouse gases, 131, 137, 137i, 138
    CO_2 and, 44, 135t, 137, 139
    methane and, 40, 109, 135t
    reducing, 42, 141, 142
Grid managers, 31, 32, 147, 152
Grids. *See* Power grids
Groundwater, 18
    geothermal power plants and, 57

H

Hahn, Otto, 26
Haze, regional, 136
Heat
    fossil fuel cycle and, 134i
    oceanic, 73, 76
    solar, 73, 76, 83, 84, 88, 92
    sources of, 144, 145, 148
Heat engines, 21
Heat exchangers, 53
    in CSP systems, 88
    in nuclear plants, 128i, 129
    in parabolic troughs, 89, 89i
Heliostats, 90, 90i
Henry, Joseph, 22
Hertz, 24
Hindenburg, 110
Hot springs, 51, 51i, 54, 57, 57i
Hoover Dam, 65
Hybrid willows, 41
Hydrocarbons, 118
Hydroelectric power, 24, 25, 61, 63.
      See also Hydropower
Hydrogen, 107–114
    as alternative fuel, 35
    automobiles, 108, 110, 112, 147
    biomass and, 109
    from coal gasification, 110
    gas combustible fuel, 111
    electrolysis and, 109–110
    as energy source, 34i
    expense of, 114
    flammability of, 113
    in fossil fuels, 114
    fuel cells and, 26, 107, 110–111,
      111i, 113
    natural gas and, 109
    pollution and, 111, 113
    portability of, 113
    producing, 108–110

    renewable/nonrenewable, 107,
      110, 114
    safety issues of, 113
    steam reforming and, 110
    usage, 112, 112i, 114
Hydrogen sulfide, 58
Hydropower, 17, 19, 24, 25, 33, 34i,
      61–70
    as energy, 24, 33, 34i, 35, 36
    large, 63, 69
    pollution from, 69
    rainfall and, 62, 62i, 69
    storage systems and, 64, 64i
    run-of-river and, 63, 64, 66–68, 69
Hydropower plants, 24, 24i
    advent of, 17, 19, 61
    capacity of, 68, 69
    combined, 66–67
    dams and, 69, 70
    design of, 62, 63, 64–65, 66–68
    electricity from, 63
    fish and, 65, 65i, 66, 69
    future of, 69, 70
    low impact of, 66
    proliferation of, 61
    run-of-river, 63, 64, 66–68, 69
    shortcomings of, 68, 69, 70
    storage, 64, 64i, 69
    water falls and, 62, 62i

I

Iceland
    geothermal in, 55, 56, 56i, 57, 57i
    hydrogen in, 113
Incandescent light bulb, 146
Industrial Revolution, 19–21
    biomass and, 39
    cities and, 21i
    fossil fuels and, 35, 117
    waterwheels in, 61
Infrared waves, 137, 137i
Intergovernmental Panel on Climate
    Change, 139
Internal combustion engines, 23
Iron production, 18, 20, 20i

K

Kilowatt (kW), defined, 30

L

La Rance, 74, 76
Landfills, 39, 41, 42, 44
    gas, 41, 45, 109

INDEX (continued)

Lava, 50
Lighting, 143, 144, 145i, 146
Light bulbs, 23
    compact fluorescent (CFL), 144,
        145i, 146
    efficiency of, 30, 115, 146, 146i
    incandescent, 146, 146i
Lightning, vii-viii, 22
Limestone, 141
Liquefied natural gas (LNG), 119, 123
Load shifting, 152
Locomotives, 21, 121
London, Jack, 55
Lower Robertson project, 68

M

Magma, 50, 50i
Magnets, 22, 28, 28i
Mammoth Lakes, 49
Mantle, 50, 50i
Manufacturing, 19, 20, 25
Manure, 30
    gasification of, 41, 45
    methane and, 44
Marine currents, 74-75
    energy systems, 76-77, 78, 78i
McCabe Wave Pump, 79
Megawatts (MW), defined, 30
Meitner, Lise, 26
Meters, 31, 32, 152
Methane. See also Natural gas
    gasification and, 40-41, 45
    hydrogen in, 109, 112
    manure and, 44
    microbes and, 40, 109, 147i
    production, 119
Michelin, André, 23
Michelin, Edouard, 23
Microbes, 40
    in hydrogen production, 109
    micro-turbines and, 147i
Microgrids, 147
Micro-turbines, 147, 147i
Minerals, 118, 126
    geothermal, 57, 58
    ocean, 73
Mining, 19, 67, 135t, 146
    coal, 18, 18i, 19, 42, 112, 117, 118,
        120, 120i, 124, 146
    uranium, 126, 129
Mississippi River, 67
Model T, 24
Mont-Saint-Michel, 74

Morse code, 22
Mouchet, Auguste, 25
Mud pots, 51
MW. See Megawatts

N

NASA
    hydrogen and, 26, 107, 110
National Oceanic and Atmospheric
    Administration (NOAA), 139
Natural gas. See also Methane
    convenience of, 20-21, 25
    formation of, 117, 119
    hydrogen and, 109
    liquefied natural gas (LNG), 119,
        123
    odor of, 119
    power plants and, 120, 121, 122,
        123
    production, 119
    shortages of, 122, 123
    transportation of, 119, 123
    in turbines, 43, 43i, 121
Net metering, 153
Neutrons, 30, 129
Niagara Falls, 24, 61
Nitrogen oxides, 135t, 136
NOAA. See National Oceanic and
    Atmospheric Administration
Nodding Duck, 78i
Nonrenewable energy, 33, 34i
Nuclear chain reaction, 127, 127i
Nuclear energy, 24, 26, 125-131. See
    also Uranium
    electricity from, 127-129, 130, 131
    pollution and, 129, 130
    safety, 129, 130, 131
Nuclear fission, 26
    history of, 26, 125
Nuclear fuels, 33, 126, 128
    as energy source, 33, 34i
    recycling of, 129, 131
    uranium and, 126, 127, 128, 129
Nuclear fusion, 129
Nuclear power plants, 26, 125i, 128,
        128i, 129, 131
    in electrolysis, 110
    Three Mile Island accident, 130
Nuclear reactions, 125, 126, 127
Nuclear Regulatory Commission, 131
Nuclear waste, 139, 131
Nucleus, 30, 125, 126, 127

O

Ocean energy, 33, 34i, 73-80, 81-82
    currents and, 74-75
    electricity from, 73, 74, 81-82
    as energy source, 33, 34i, 73, 76
        77, 80
    heat and, 76, 76i
    pollution and, 80
    power plants and, 77i, 79, 80
    thermal energy, 79, 79i, 80
    tidal currents, 74-75
    waves and, 75, 77, 78, 78i
Ocean Thermal Energy Conversion
        (OTEC), 79, 79i, 80
    closed cycle, 79
    efficiency of, 80
    open cycle, 79, 79i
Øersted, Hans, 22
Oil, 22, 117. See also Gasoline
    convenience of, 23, 24, 28
    crude, 118-119, 119i
    derricks, 22i
    history of, 22-23, 24
    production, 118-119
    shortages of, 35, 122
    spills, 123
Oklo mine, 126
OTEC. See Ocean Thermal Energy
    Conversion

P

Parabolic troughs, 88, 89, 89i
Paris Exhibition, 25
Parsons, Charles, 23
Passive solar, 17, 84, 148. See also
    Solar energy
Peak load, 152
Peaking power, 32
    geothermal and, 58
    hydropower and, 69
    natural gas and, 122, 124
    tidal and, 80
    wind and, 101
Penstock, 63
Petroleum. See Oil
Photoelectric effect, 86
Photons, 83, 85
Photosynthesis, 134i, 141
Photovoltaics (PV), 84-87 See also
    Solar energy
    arrays, 85
    cells, 86, 86i
    grid-connected, 85, 87

improvements in, 26, 83
kits (Solar Home Systems), 87
manufacturing, 92, 91
panels, 85, 85i, 91
stand-alone, 85, 86-87, 92
thin film, 84, 85
Pifre, Abel, 25
Pinnacles National Monument, 86, 86i
Plutonium, 126, 131
Policies
for energy, 150-151
taxes and, 155
Pollution, 133, 135t, 136, 153. *See also* Air pollution
biomass and, 39
causes of, 39
coal and, 121
cogeneration and, 149
controls, 39, 151
fossil fuels and, 118, 121, 123
gas turbines and, 121
hydrogen and, 114
hydropower and, 69
nuclear and, 129, 130, 131
ocean power and, 80
solar and, 91
solutions to, 138-139, 142, 144, 151
wind and, 97, 101
Pollution control technology, 138
Poplar trees, 41
Power grids, 31, 32, 147, 152
Power plants, 23, 24, 27, 28. *See also* specific power plant types
biomass v. fossil fuel, 43, 44
capacity of, 35
coal, 24, 120-121, 124
cofiring, 44
cogeneration and, 149
combined-cycle, 121, 122
diesel, 122, 124
distributed generation (DG), 147
efficiency of, 146, 146i, 147
forced-air, 26, 77, 78i
fossil fuel, 117-119, 122
natural gas and, 123
nuclear, 10, 98-99, 98i
OTEC and, 79, 79i, 80
pollution from, 123
solar and, 27, 90, 90i, 91, 92,
steam-driven, 27, 27i, 28, 29, 29i
voltage and, 31
waste-to-energy, 42, 45, 46
wave, 77-78, 80

Power producers, regulating, 150
Power towers, 90
Precipitation, 62, 62i, 140
Pricing, 152-155
Production Tax Credit, 155
Propane, 107i, 119
Protons, 30
Pumping rig, 119i
PV. *See* Photovoltaics

Q

Quito (Ecuador), 63

R

Radiation, 83, 84, 126, 137, 137i
Radioactive elements, 50
Radioactivity, 126
Rainfall, 51, 62, 62i, 140
Rainforests, 141, 142
Rebates, 92, 153
RECs (credits), 154
Recycling, 42
nuclear fuel and, 129, 131
Refrigerators, 144, 146, 148
Relativity, concept of, 24
Renewable energy, 33, 34i, 35
biomass, 33, 34i, 39-46
credits (RECs), 154
definition of, 33
geothermal, 33, 34i, 49-58
hydropower, 33, 34i, 61-70
ocean, 33, 34i, 73-80
resources, 34i
solar, 33, 34i, 83-92
strategies, 154-155
tradeable renewable energy credits (TRECs), 154
wind, 33, 34i, 95-101
Renewable portfolio standards, 154
Research, government, 155
Restoration ecology, 142
Ring of Fire, 56, 56i
Roads, 25
Roosevelt, Teddy (Theodore), 55
RPS. *See* Renewable portfolio standards

S

Sacramento Municipal Utility District, 87
Sahara Desert, 141, 148
Salton Sea, 55
Savery, Thomas, 18

Schatz Solar Hydrogen Project, 108
Scrubbers, 123
Separators, geothermal, 52, 52i, 53
Shannon River Estuary, 79
Sierra Nevada mountain range, 49, 63
Silica, 58
Silicon, 86
Smart grids, 147
Smart meters, 32
Smelting, 20
Smog, photochemical, 21, 136
Solar cells. *See* Photovoltaics
Solar collectors, 148
Solar dish engines, 88, 88i, 92
Stirling, Robert, and, 21, 88
Solar energy, 26, 33, 34i, 83-92, 93-94. *See also* Photovoltaics (PV); Solar power
central receiving towers and, 90, 90i
concentrating solar power (CSP), 92
dish engines and, 88, 88i
electricity and, 84-90, 93-94
as energy source, 33, 34i
Greeks and, 17
kits and, 87
parabolic troughs and, 89, 89i
power plants and, 27, 90, 90i, 91, 92
PV and, 84-87
in steam engines, 21, 25, 25i
solar thermal and, 88-90
storing, 88, 92
systems, 91
Solar heating, 145, 148
Solar Home Systems, 87
Solar ponds, 88
Solar power. *See also* Solar energy
active solar, 84, 148
developments in, 26
passive form, 17, 148
plants, 27, 90, 90i, 91, 92
Solar radiation, 76, 83i, 137, 137i
Solar resource, 83, 84-85
Solar Spectrum, 84
Solar technology, cost of, 91
Solar thermal plants, 88-90, 92
Soleil Journal, 25
Stanley, William, 24
Steam
in power plants, 27, 27i, 29, 29i
reforming, 110
water and, 29, 29i

INDEX (continued)

Steam engines, 18i, 19i, 21, 23
   coal in, 18-19
   in cotton mills, 19i
   history of, 18, 20
   in Industrial Revolution, 19-21
   solar, 25, 25i
Steamboats
   paddle-wheel, 20
   Turbinia and, 23, 23i
Steel, 20
Stirling engine, 21, 88, 88i
Stirling, Robert, 21
Strip mining, 118, 124
Subatomic particles, 125-126, 127
Sulfur, 20, 58, 118, 135t
Sulfur oxides, 134i, 135t, 136
Sun, 17, 83, 86i
Sustainable energy, 35
Switchgrass, 41, 41i
Syngas, 109

T

Tax policies, 155
Tazimina Hydroelectric Project, 67
Tectonic plates, 50, 50i
Telegraph, 22
Tennessee Valley Authority (TVA), 65
Tesla, Nikola, 23, 24
Textile industry, 19
Thermal energy (ocean), 79, 79i, 80
Three Gorges project, 69
Three Mile Island accident, 130
Tidal power, 76, 77, 77i, 80
Tidal power plants, 77, 77i, 79, 79i, 80
   La Rance, 74, 76
Time of use, 152
Tires, 23
Town gas, 20-21, 107
Tradeable renewable energy credits
   (TRECs), 154
Trains, 21, 120, 121
Transformers, 31
Transmission lines, 31
Transportation, 17
   fuels and, 19-20, 21, 22-23, 26, 35
   hydrogen in, 108, 110, 112, 147
   urban, 23, 24
TRECs (credits), 154
Trees
   deciduous, 145, 145i, 148
   energy farms and, 41
   fast-growing, 39, 39i
   as renewable resources, 155

Trimmings (biomass), 39, 42, 44, 109
Trolley lines, 24
Turbines, 28, 31. *See also* Wind
   turbines
   conventional gas, 121
   gas, 43, 43i, 121
   geothermal, 49, 52, 52i, 53, 53i
   improvements in, 121, 121i
   steam-driven, 27, 27i, 29
   sub-sea, 76
   water, 61, 77
Turbinia, 23, 23i
20,000 Leagues Under the Sea, 80

U

Ultraviolet rays, 83, 84
Uranium, 126, 127. *See also* Nuclear
   energy
   characteristics of, 126
   enrichment of, 128, 129
   in nuclear power plants, 26, 125i,
     128, 128i, 129, 131
   reactivity of, 126, 127, 127i
   safety, 129
   U-235 and, 126, 127i, 128
   weight of, 126, 127
U.S. Bureau of Reclamation, 65
U.S. Congress, 151
U.S. Department of Defense, 112
U.S. Department of Energy, 99, 155
U.S. Environmental Protection Agency
   (EPA)
   Clean Air Act and, 151
   ENERGY STAR program of, 153
U.S. Marine Corps base, 78
U.S. Navy, 125
U.S. Treasury, 155

V

Valdez oil spill, 123
Verne, Jules, 80
Volcanoes, 49, 56, 56i, 135t
Volta, Alessandro, 22, 30
Voltage, 30, 21
Volts, 30

W

Waste-to-energy power plants, 42,
   45, 46
Water, 17, 62, 62i
   boiling point of, 29
   cycle of, 62, 62i
   drinking, 79

   in electrolysis, 108
   falls, 62, 62i
   geothermal, 17, 49, 149
   global warming and, 139, 140
   head of, 62
   heaters, 144, 145
   hydrogen in, 107
   in power plants, 28, 29, 29i
   pumps, 18
   reservoirs and, 51, 51i, 58
   steam and, 19
   in steam engines, 18, 18i, 19, 19i,
     20-21, 23
Water falls, 62, 62i
Watermills, 17, 28
Waterwheels, 17, 19, 28, 61
   as turbines, 28
   v. water turbine, 63
Watt, James, 20-21, 30
Watt-hours (Wh), 30, 31, 32
Watts (W), 30
Wave energy, 75, 76-78
Wave Energy Devices, 77-78, 78i
Waves, 75
Westinghouse, George, 24
Wet-cell battery, 22
Wind energy, 17, 33, 34i, 95-101,
   102-103
   anemometer, 96, 96i
   classification of, 26, 95
   developments in, 95
   electricity from, 97, 98, 99-100,
     101, 102-103
   history of, 17, 95
   storage of, 97, 101
Wind farms, 97, 98-99, 144
Wind turbines, 96, 96i, 97, 98, 99
   birds and, 99, 101
   in cities, 100
   efficiency of, 101
   noise and, 101
   sizes, 99-100
   stand-alone, 97
Windmills, 17, 95
Wood, 17-18, 20, 39

XYZ

Yangtze project, 65
Zinc, 5

THE AUTHORS

Marilyn L. Nemzer, M.A., Executive Director of the Energy Education Group, the Geothermal Education Office and the Global Book Exchange, is the developer, co-author and editor of *Energy for Keeps*. Nemzer, with an early background in teaching and public education policy, has been collaboratively producing award-winning energy education projects and materials for 20 years. She serves on advisory boards of several state, national, and international organizations that focus on renewable energy and the environment and is a trustee of the Marin County Board of Education.

Deborah S. Page, M.A., is the lead writer and co-author of *Energy for Keeps*. Page, a classroom teacher, reading specialist, and science educator with the Claremont Unified School District in California, and principal of Page One Productions, is a consummate researcher with abiding interest in environmental issues. She has over a decade of experience writing reader-friendly semi-technical materials on renewable energy and energy conservation and has been an invited course developer and instructor at the Yellowstone Institute.

Anna K. Carter, co-author and technical editor of *Energy for Keeps*, is a consultant who has worked in the renewable energy field for over 25 years, specializing in energy policy, regulatory compliance, project permitting, and public information. She has authored and co-authored numerous articles and renewable energy public information materials and served for many years on the Board of Directors of the Geothermal Resources Council, a renewable energy educational organization with over 1,100 U.S. and International members.

This talented team has worked together for many years to create quality energy education materials for both the general public and the classroom. They are committed to high standards of accuracy and inclusion. Throughout the years of research and writing devoted to *Energy for Keeps*, the team was assisted by over 75 experts in education, energy, engineering, and electricity generation

To order more copies of this book:

Ordering information at **www.energyforkeeps.org**
Quantity discounts and supplementary teaching materials available

Or contact:

Marilyn Nemzer
Director
Energy Education Group
664 Hilary Drive
Tiburon, CA 94920

tel: 415.435.4574

email: energyforkeeps@aol.com

ENERGY
EDUCATION
GROUP